Springer Finance

SO-BZL-112

Springer

Berlin
Heidelberg
New York
Barcelona
Hong Kong
London
Milan
Paris
Tokyo

Springer Finance

Springer Finance is a new programme of books aimed at students, academics and practitioners working on increasingly technical approaches to the analysis of financial markets. It aims to cover a variety of topics, not only mathematical finance but foreign exchanges, term structure, risk management, portfolio theory, equity derivatives, and financial economics.

Credit Risk: Modeling, Valuation and Hedging
T. R. Bielecki and M. Rutkowski
ISBN 3-540-67593-0 (2001)

Risk-Neutral Valuation: Pricing and Hedging of Financial Derivatives
N. H. Bingham and R. Kiesel
ISBN 1-85233-001-5 (1998)

Interest Rate Models – Theory and Practice
D. Brigo and F. Mercurio
ISBN 3-540-41772-9 (2001)

Visual Explorations in Finance with Self-Organizing Maps
G. Deboeck and T. Kohonen (Editors)
ISBN 3-540-76266-3 (1998)

Mathematics of Financial Markets
R. J. Elliott and P. E. Kopp
ISBN 0-387-98553-0 (1999)

Mathematical Finance – Bachelier Congress 2000
H. Geman, D. Madan, S. R. Pliska and T. Vorst (Editors)
ISBN 3-540-67781-X (2001)

Mathematical Models of Financial Derivatives
Y.-K. Kwok
ISBN 981-3083-25-5 (1998), second edition due 2001

Efficient Methods for Valuing Interest Rate Derivatives
A. Pelsser
ISBN 1-85233-304-9 (2000)

Marc Yor

Exponential Functionals of Brownian Motion and Related Processes

 Springer

Marc Yor
Université de Paris VI
Laboratoire de Probabilités et Modèles Aléatoires
175, rue du Chevaleret
75013 Paris
France

Translation from the French of chapters [1], [3], [4], [8]
Stephen S. Wilson
Scientific Translator
Technical Translation Services
31 Harp Hill
Cheltenham GL52 6PY
Great Britain

ssw@stephenswilson.co.uk
http://www.techtrans.cwc.net

Library of Congress Cataloging-in-Publication Data
Yor, Marc.
Exponential functionals of Brownian motion and related processes / Marc Yor. p. cm. –
(Springer finance)
Includes bibliographical references.
ISBN 3540659439 (pbk.: alk. paper)
1. Business mathematics. 2. Finance-Mathematical models. 3. Brownian motion processes. I. Title.
II. Series.
HF5691 .Y67 2001
519.2'33--dc21   2001020860

Mathematics Subject Classification (2000): 60J65, 60J60

ISBN 3-540-65943-9 Springer-Verlag Berlin Heidelberg New York

This work is subject to copyright. All rights are reserved, whether the whole or part of the material is
concerned, specifically the rights of translation, reprinting, reuse of illustrations, recitation, broad-
casting, reproduction on microfilm or in any other way, and storage in data banks. Duplication of
this publication or parts thereof is permitted only under the provisions of the German Copyright
Law of September 9, 1965, in its current version, and permission for use must always be obtained
from Springer-Verlag. Violations are liable for prosecution under the German Copyright Law.

Springer-Verlag Berlin Heidelberg New York
a member of BertelsmannSpringer Science+Business Media GmbH

http://www.springer.de

© Springer-Verlag Berlin Heidelberg 2001
Printed in Germany

The use of general descriptive names, registered names, trademarks, etc. in this publication does not
imply, even in the absence of a specific statement, that such names are exempt from the relevant pro-
tective laws and regulations and therefore free for general use.

Typeset in TₑX by Hindustan Book Agency, New Delhi, India, except for chapters 1, 3, 4, 8 (typeset by
Stephen S. Wilson)

Cover design: *design & production* GmbH, Heidelberg

Printed on acid-free paper   SPIN 10730209   41/3142LK – 5 4 3 2 1 0

Preface

This monograph contains:

    - ten papers written by the author, and co-authors, between December 1988 and October 1998 about certain exponential functionals of Brownian motion and related processes, which have been, and still are, of interest, during at least the last decade, to researchers in Mathematical finance;

    - an introduction to the subject from the view point of Mathematical Finance by H. Geman.

The origin of my interest in the study of exponentials of Brownian motion in relation with mathematical finance is the question, first asked to me by S. Jacka in Warwick in December 1988, and later by M. Chesney in Geneva, and H. Geman in Paris, to compute the price of Asian options, i.e.: to give, as much as possible, an explicit expression for:

$$C^{(\nu)}(t,k) \stackrel{\text{def}}{=} E\left[\left(\frac{1}{t}A_t^{(\nu)} - k\right)^+\right] \tag{1}$$

where $A_t^{(\nu)} = \int_0^t ds \exp 2(B_s + \nu s)$, with $(B_s, s \geq 0)$ a real-valued Brownian motion.

Since the exponential process of Brownian motion with drift, usually called: geometric Brownian motion, may be represented as:

$$\exp(B_t + \nu t) = R_{A_t^{(\nu)}}^{(\nu)}, \qquad t \geq 0, \tag{2}$$

where $(R_u^{(\nu)}, u \geq 0)$ denotes a δ-dimensional Bessel process, with $\delta = 2(\nu+1)$, it seemed clear that, starting from (2) [which is analogous to Feller's representation of a linear diffusion X in terms of Brownian motion, via the scale function and the speed measure of X], it should be possible to compute quantities related to (1), in particular:

$$\int_0^\infty dt\, e^{-\lambda t} E[(A_t^{(\nu)} - k)^+]$$

in hinging on former computations for Bessel processes. This program has been carried out with H. Geman in [5] (a summary is presented in the C.R.A.S. Note [3]).

    As a by-product of this approach, the distribution of $A_{T_\lambda}^{(\nu)}$ (i.e.: the process $(A_t^{(\nu)}, t \geq 0)$ taken at an independent exponential time T_λ with parameter λ), was obtained in [2] and [4]: the distribution of $A_{T_\lambda}^{(\nu)}$ is that of the ratio of a beta variable, divided by an independent gamma variable, the parameters of which depend (obviously) on ν and λ. When ν is negative, it is also natural to consider $A_\infty^{(\nu)}$ which, as proved originally by D. Dufresne, is distributed as the reciprocal of a gamma variable; again, the representation (2) and known results on Bessel processes give a quick access to this result.

An attempt to understand better the above mentioned ratio representation of $A_{T_\lambda}^{(\nu)}$ is presented in [6], along with some other questions and extensions.

My interest in Bessel processes themselves originated from questions related to the study of the winding number process $(\theta_t, t \geq 0)$ of planar Brownian motion $(Z_t, t \geq 0)$, which may be represented as:

$$\theta_t = \gamma \left(\int_0^t \frac{ds}{|Z_s|^2} \right), \qquad t \geq 0, \tag{3}$$

where $(\gamma(u), u \geq 0)$ is a real-valued Brownian, independent of $(|Z_s|, s \geq 0)$. The interrelations between planar Brownian motion, Bessel processes and exponential functionals are discussed in [7], together with a comparison of computations done partly using excursion theory, with those of De Schepper, Goovaerts, Delbaen and Kaas in vol. 11, n° 4 of *Insurance Mathematics and Economics*, done essentially via the Feynman - Kac formula.

The methodology developed in [2], [3], [4] and [5] to compute the distribution of exponential functionals of Brownian motion adapts easily when Brownian motion is replaced by a certain class of Lévy processes.

This hinges on a bijection, introduced by Lamperti, between exponentials of Lévy processes and semi-stable Markov processes.

A number of computational problems remain in this area; some results about the law of:

$$Z \stackrel{\text{def}}{=} \int_0^\infty dt \exp \left(-\xi \int_0^t ds (R_s^{(\nu)})^\gamma \right) \tag{4}$$

have been obtained in [5] and [9] (see also, in the same volume of Mathematical Finance, the article by F. Delbaen: *Consols in the C.I.R. model*).

It is my hope that the methods developed in this set of papers may prove useful in studying other models in Mathematical Finance.

In particular, models with jumps, involving exponentials of Lévy processes keep being developed intensively, and I should cite here papers by Paulsen, Nilsen and Hove, among many others; see, e.g., the references in [A].

Concerning the different aspects of studies of exponential functionals, D. Dufresne [B] presents a fairly wide panorama.

An effort to present in a unified manner the methodology used in some of the papers in this Monograph is made in [C].

To facilitate the reader's access to the bibliography about exponential functionals of Brownian motion, I have:

a) systematically replaced in the references of each paper/chapter of the volume the references "to appear" by the correct, final reference of the published paper, when this is the case;

b) added at the end of (each) chapter $\#N$, a Postscript $\#N$, which indicates
    some progress made since the publication of the paper, further references,
    etc. . .

Finally, it is a pleasure to thank the coauthors of the papers which are
gathered in this book; particular thanks go to H. Geman whose persistence
in raising questions about exotic options, and more generally many problems
arising in mathematical finance gave me a lot of stimulus.

Last but not least, special thanks to F. Petit for her computational skills
and for helping me with the galley proofs.

References

[A] Carmona, P., Petit, F. and Yor, M. (2001) Exponential functionals of Lévy pro-
     cesses. *Birkhäuser volume: "Lévy processes: theory and applications"* edited by:
     O. Barndorff-Nielsen, T. Mikosch, and S. Resnick, p. 41–56.
[B] Dufresne, D. Laguerre series for Asian and other options. *Math. Finance*, vol. **10**,
     n° 4, October 2000, 407–428
[C] Chesney, M., Geman, H., Jeanblanc-Picqué, M., and Yor, M. (1997). Some Com-
     binations of Asian, Parisian and Barrier Options. *In: Mathematics of Derivative
     Securities,* eds: M.A.H. Dempster, S.R. Pliska, 61–87. Publications of the New-
     ton Institute. Cambridge University Press

Table of Contents

Preface v

0. Functionals of Brownian Motion in Finance
   and in Insurance 1
   by Hélyette Geman

1. On Certain Exponential Functionals of
   Real-Valued Brownian Motion 14
   J. Appl. Prob. **29** *(1992), 202–208*

2. On Some Exponential Functionals of Brownian Motion .. 23
   Adv. Appl. Prob. **24** *(1992), 509–531*

3. Some Relations between Bessel Processes, Asian
   Options and Confluent Hypergeometric Functions 49
   C.R. Acad. Sci., Paris, Sér. I **314** *(1992), 471–474*
   (with Hélyette Geman)

4. The Laws of Exponential Functionals of Brownian
   Motion, Taken at Various Random Times 55
   C.R. Acad. Sci., Paris, Sér. I **314** *(1992), 951–956*

5. Bessel Processes, Asian Options, and Perpetuities 63
   Mathematical Finance, Vol. **3**, *No. 4 (October 1993), 349–375*
   (with Hélyette Geman)

6. Further Results on Exponential Functionals
   of Brownian Motion 93

7. From Planar Brownian Windings to Asian Options 123
   Insurance: Mathematics and Economics **13** *(1993), 23–34*

8. On Exponential Functionals of Certain Lévy Processes .. 139
   Stochastics and Stochastic Rep. **47** *(1994), 71–101*
   (with P. Carmona and F. Petit)

9. On Some Exponential-integral Functionals
   of Bessel Processes 172
   Mathematical Finance, Vol. **3**, *No. 2 (April 1993), 231–240*

10. Exponential Functionals of Brownian Motion
    and Disordered Systems 182
    J. App. Prob. **35** *(1998), 255–271*
    (with A. Comtet and C. Monthus)

Index 205

0
Functionals of Brownian Motion in Finance and in Insurance

by Hélyette Geman
University Paris-Dauphine and ESSEC

1. Introduction

In 1900, the mathematician Louis Bachelier proposed in his dissertation "Théorie de la Spéculation" to model the dynamics of stock prices as an arithmetic Brownian motion (the mathematical definition of Brownian motion had not yet been given by N. Wiener) and provided for the first time the exact definition of an option as a financial instrument fully described by its terminal value. In his 1965 paper "Theory of Rational Warrant Pricing", the economist and Nobel prize winner Paul Samuelson, giving full recognition to Bachelier's fondamental contribution, transformed the arithmetic Brownian motion into a geometric Brownian motion assumption to account for the fact that stock prices cannot take negative values. Since that seminal paper – in the appendix of which H.P. McKean provided a closed-form solution for the price of an American option with infinite maturity-, the stochastic differential equation satisfied by the stock price

$$dS_t = \mu S_t dt + \sigma S_t d\hat{W}_t \tag{1}$$

has been the central reference in the vast number of papers dedicated to option pricing. It was in particular the assumption also made by Black-Scholes (1973) and Merton (1973) in the papers providing the price at date t of a European call paying at maturity T the amount $\max(0, S_T - k)$.

Later on, options were introduced whose payout depends not only on S_T but also on the values of S_t over the whole interval $[0, T]$, hence the qualification "path-dependent" given to these options. These instruments, barrier, average or lookback, involve quantities such as the maximum, the minimum or the average of S_t over the period. Consequently, their prices involve functionals of the Brownian motion present in equation (1). As will be shown in this chapter and, more generally throughout this book, the knowledge of the mathematical properties of functionals of Brownian motion plays a crucial role for conducting the computations involved in their valuation and exhibiting closed-form or quasi closed-form results.

Over the last ten years, path-dependent options have become increasingly popular in equity markets, and even more so in commodity and *FX* markets. As of today, ninety five per cent of options exchanged on oil and oil spreads are Asian. On the other hand, barrier options allow portfolio managers to hedge at a lower cost against extreme moves of stock or currency prices.

In order to overcome the technical difficulties associated with the valuation and hedging of path-dependent options even in the classical geometric Brownian motion setting of the Black-Scholes-Merton model, practitioners taking advantage of the power of new computers and workstations make with good reasons a great use of Monte Carlo simulations to price path-dependent options. However, our claim is that the results are not always extremely accurate: the most obvious example is the case of "continuously (de)activating" barrier options, heavily traded in the *FX* markets, and where the option is activated (or desactivated) at any point in the day where the underlying exchange rate hits a barrier. We recall that a barrier option provides the standard Black-Scholes payoff max $(0, S_T - k)$ only if (or unless) the underlying asset S has reached a prespecified barrier L, smaller or greater than the strike price k, during the lifetime $[0, T]$ of the option. In the equity markets, the classical situation for barrier contracts is that S_t is compared with L only at the end of each day (daily fixings). In contrast, in the *FX* markets, the comparison takes place quasi-continuously and (de)activation may occur at any point in time. Obviously, the valuation of the option by Monte Carlo simulations built piecewise may lead to fatal inaccuracies, in particular when the value of the underlying instrument is near the barrier close to maturity (entailing at the same time hedging difficulties well-known by option traders).

Along the same lines, when computing the Value at Risk (VaR) of a complex position or of a portfolio (we recall that the value at risk for a given horizon T and a confidence level p is the maximum loss which can take place with a probability no larger than p), Monte Carlo simulations allow one to represent different scenarios on the state variables. But if the price of every exotic security in each scenario is in turn computed through Monte Carlo simulations, one has to face "Monte Carlo of Monte Carlo" and it becomes impossible, even with powerful computers, to calculate the VaR of the portfolio overnight. In an analytical approach, since we obtain explicit or quasi-explicit solutions, the new values of the options can immediately be computed by incorporating in the pricing formulas the parameters corresponding to the different scenarios; hence, the problems mentioned above in estimating VaR are dramatically reduced. The remainder of this chapter is organized as follows. Section 2 recalls the definition of stochastic time changes and shows why they are very useful to price (and hedge), via Laplace transforms, path-dependent options. Section 3 examines the specific case of Asian options and offers comparisons with Monte Carlo simulation prices. Section 4 addresses the case of barrier and double-barrier options and illuminates the hedging difficulties near maturity when the underlying asset price is close to a barrier. Section 5 contains some concluding remarks.

2. Stochastic Time Changes and Laplace Transforms

Representing the randomness of the economy by the filtered probability space (Ω, F, F_t, P) where F_t represents (the filtration of) information available at time t and P the objective probability measure, we assume, as in the classical Black-Scholes-Merton setting, that the dynamics of the underlying asset price process are driven by the stochastic differential equation

$$\frac{dS_t}{S_t} = \mu dt + \sigma d\hat{W}_t \tag{1}$$

where μ is a real number, σ is a strictly positive number and $(\hat{W}_t)_{t \geq 0}$ is a P-Brownian motion. Introducing the assumption of no arbitrage, we know from the seminal papers by Harrison-Kreps (1979) and Harrison-Pliska (1981), that there exists a so-called risk adjusted probability measure Q under which the dynamics of S_t become

$$\frac{dS_t}{S_t} = (r - y)dt + \sigma dW_t \tag{1'}$$

where $(W_t)_{t \geq 0}$ is a Q-Brownian motion and y denotes the continuous dividend rate of the underlying stock, supposed to be constant over the lifetime $[0, T]$ of this option. Equation (1) expresses the key mathematical assumption in the Black-Scholes model. From an economic standpoint, since there is in this representation one source of risk, namely the Brownian motion $(\hat{W}_t)_{t \geq 0}$, it follows from central results of finance such as the Capital Asset Pricing Model or the Arbitrage Pricing Theory (S. Ross, 1976) that the expected return on a risky security should outperform the risk-free rate r over the period by *one* risk premium, i.e.,

$$E_P \left(\frac{dS_t}{S_t} \right) = dt(r + \textit{risk premium})$$

where, for simplicity, r is supposed to be constant in the Black-Scholes model. The risk premium can be written as the positive constant σ times a quantity λ (called the market price of equity risk). It is then possible to rewrite equation (1) as

$$\frac{dS_t}{S_t} = rdt + \sigma(\lambda dt + d\hat{W}_t)$$

and Girsanov's theorem allows to obtain equation (1') (it also provides the expression of the Radon-Nikodym derivative $\frac{dQ}{dP}$ in terms of λ - whether the risk premium parameter is supposed to be constant or not – and $(\hat{W}_t)$). From now on, we will be working under the probability measure Q in order to be allowed to write the price of an option as the expectation of the discounted terminal pay-off.

As explained for instance by Kemna and Vorst (1990) who studied the valuation of average-rate options when these instruments started becoming very popular in the financial markets, the pricing difficulties are fundamentally conveyed by the fact that the representation of $(S_t)_{t\geq0}$ by a geometric Brownian motion as described in equation (1) (and which is crucial for the simple proofs-through a partial differential equation or a probabilistic approach-of the Black-Scholes formula) is not transmitted to the average of S; hence, the idea - developed in Geman-Yor (1993) − of searching for a class of stochastic processes stable under additivity and related to the geometric Brownian motion. The so-called squared-Bessel processes, denoted hereafter $(BESQ(u), u \geq 0)$ have the remarkable property of being Markov processes (which is the assumption common to nearly all models of option pricing and yield curve deformations) and of being stable by additivity. Moreover, a particular case of a remarkable theorem about the exponentiation of Lévy processes, due to Lamperti (1972) establishes that

$$S(t) = BESQ[X(t)] \tag{2}$$

where the processes X and $BESQ$ are completely defined in terms of the parameters of the geometric Brownian motion $(S(t))_{t\geq0}$. Equation (2) simply states that the value of S at time t is equal to the value of the squared-Bessel process $BESQ$ at time $X(t)$. X defines a *stochastic time change* and formula (2) expresses that a geometric Brownian motion is a time-changed squared-Bessel process. The main condition a process X has to satisfy in order to define a time change is to be (almost surely) increasing since time cannot go backwards; moreover, the $(X(t), t \geq 0)$ have to be stopping times relative to an appropriate filtration. Other properties such as independent or identical increments may or may not be satisfied; when both are satisfied, the time change is called subordinator (see Bochner (1955)). Stochastic time changes are very useful for the pricing of exotic options, as this chapter will try to show. They have also become extremely popular when studying asset price dynamics: $X(t)$ may represent random sampling times in a financial time series as a function of calendar time t. The time change $X(t)$ may also account for differences in market activity at different hours in the day or because of new information release: Ané-Geman (2000), analysing equity indexes and individual stocks, show that an appropriate stochastic clock allows to recover a quasi-perfect normality for asset returns.

Coming back to exotic options (barrier, double-barrier or corridor options) pricing, and assuming that the underlying asset return (S_t) is a geometric Brownian motion with drift, the quantities whose expectations (under the right probability measure) provide the option prices involve functionals of $(S_t, t \leq T)$ such as its maximum M_T or its minimum I_T over the period $[0, T]$. The trivariate joint distribution of (M_T, I_T, S_T) has been known for some time (see Bachelier 1941); however, its expression is complex. Pitman-Yor (1992) show (see also Revuz-Yor, 1998, p. 509) that this quantity becomes much simpler when the fixed time T (maturity of the option in our setting) is replaced by an exponential time τ independent of the Brownian motion contained in the dynamics of the process $(S_t)_{t\geq0}$ (this quantity is also simpler when T is infinite, a property which is consistent with the fact that since McKean's finding in 1965 on the valuation of a perpetual American option, no closed-form solution has been exhibited as of today for the finite maturity case). Remembering the expression of the density of an exponential variable, it is easy to see that in order to exploit the above mentioned property, we are naturally led to compute the Laplace transform of the option price with respect to its maturity since $\int_0^{+\infty} C(T)e^{-\lambda T}dT$ can be interpreted (up to the factor λ) as the expectation of $C(\tau)$ where τ is an exponential variable. Lastly, let us recall that the integral $\int_0^\infty S(s)ds$ is distributed as an inverse gamma variable. This interesting result was first proved by Dufresne (1990) in the analysis of perpetuities; another proof was given by Yor (1992). It is important to observe however that the integral only converges if $r-y < \frac{\sigma^2}{2}$ (hence may not exist for non-dividend paying assets since $\sigma = 0.02$ already represents a fairly high volatility). Moreover, options traded in the financial markets have a finite maturity and the above described property cannot solve the Asian option valuation problem. The price $C(T)$ itself would have a much simpler expression for T infinite, and this may be valuable for life-insurance products.

3. The Case of Asian Options

As has been mentioned earlier and is substantiated by the continuously growing literature on the subject, Asian options have a number of attractive properties as financial instruments: for thinly traded assets and commodities (e.g., gold) or newly established exchanges, the averaging feature allows one to prevent possible manipulations on maturity day by investors or institutions holding large positions in the underlying asset. They are very popular among corporate treasurers who can hedge a series of cashflows denominated in a foreign currency by using an average-rate option as opposed to a portfolio of standard options; the hedge is obviously not identical but may be viewed as sufficient. Many domestic rates used in Europe and in the US as reference rates in floating-rate notes or interest rate swaps are defined as averages of spot rates; hence, caps and floors written on these rates are, by definition, Asian. To give an example very relevant in corporate finance, we can also mention the so-called *contingent value rights*: suppose a firm A wants to acquire a firm B. A is not willing to pay too high a price for the shares of company B but knows that this may lead to a failure of the takeover. Hence firm A will offer the shareholders of company B a share of the new firm AB accompanied by a contingent-value right on firm AB, maturing at time T (say two years later). This contingent-value right is nothing but an Asian put option. The put provides the classical protection of portfolio insurance; the Asian feature protects firm A for an exceptionally low market price of the share AB on day T, as well as the shareholders B in the case of a very high market price that day. These contingent-value rights were used when Dow Chemical acquired Marion Laboratory, when the French firm Rhône-Poulenc acquired the American firm Rorer and more recently, when the insurance company Axa merged with Union des Assurances de Paris to form the second largest insurance company in the world (in the last case, the corresponding contingent value rights are still trading today). To give a last example of the usefulness of Asian options, we can mention that options written on oil or on oil spreads are mostly Asian since oil indices are generally defined as arithmetic averages. Many options written on gas have the same feature and the deregulation of the gas industry worldwide has entailed a significant growth of the gas derivatives market. Let us now turn to the valuation of these instruments.

The Mathematical Setting

We assume the asset price driven under the risk-adjusted probability measure Q by the dynamics described in (1')

$$dS(t) = rS(t)dt + \sigma S(t)dW(t)$$

We also assume that the number of values whose average is computed is large enough to allow the representation of the average $A(T)$ over $[0, T]$ by the integral

$$A(T) = \frac{1}{T} \int_0^T S(u)du$$

The value of an Asian call option at time t is expressed, by arbitrage arguments, as

$$C(t) = E_Q[e^{-r(T-t)} \max(A(T) - k, 0)/F_t] \qquad (3)$$

where k is the strike price of the option and the discount factor may be pulled out of the expectation since we assumed constant interest rates. We know that the option has a unique price: there is only one source of randomness represented by the Brownian motion and a money-market instrument traded together with the risky security, which implies market completeness.

As mentioned earlier, the mathematical difficulty in formula (3) stems from the fact that, denoting $A(t) = \frac{1}{t} \int_0^t S(u)du$, the process $(A(t))_{t \geq 0}$ is *not* a geometric Brownian motion. Many practitioners (see for instance Lévy (1992)) make this simplifying assumption and can then recover a Black-Scholes type pricing formula through the mere computation of the first two moments of $A(t)$. But to our knowledge, no upper bound of the error due to this approximation was ever provided. Tight bounds for the Asian option price, however, can be found in Rogers and Shi (1995).

Let us first observe that, when the option is traded at a date t posterior to date 0, the values of the underlying asset between 0 and t are fully known; the only randomness resides in the values to be taken by S between t and T. Hence, if the values observed between 0 and t are high enough, it may already be known at time t that the Asian call option will finish in the money and that we can write

$$\max(A(T) - k, 0) = A(T) - k \quad \text{since} \quad A(T) > \frac{1}{T}\int_0^t S(u)du > k$$

Decomposing $A(T) = \frac{1}{T}\int_0^t S(s)ds + \frac{1}{T}\int_t^T S(s)ds$ and observing that the first term is fully known at date t, we obtain

$$E_Q[A(T) - k|F_t] = \frac{1}{T}\int_0^t S(s)ds - k + E_Q\left[\frac{1}{T}\int_t^T S(s)ds|F_t\right]$$

The linearity of the integral and expectation operators and the martingale property satisfied by the discounted price of S_t under Q allow one to compute explicitly the last term, namely

$$E_Q\left[\frac{1}{T}\int_t^T S(s)ds|F_t\right] = \frac{1}{T}\int_t^T E_Q[S(s)|F_t]ds$$

$$= \frac{1}{T}\int_t^T S(t)e^{r(s-T)}ds = \frac{S(t)}{rT}(1 - e^{r(t-T)})$$

We then obtain the Asian call price (when it is known at date t that the call is in the money) as

$$C(t) = S(t)\frac{1 - e^{-r(T-t)}}{rT} - e^{-r(T-t)}\left[k - \frac{1}{T}\int_0^t S(s)ds\right] \qquad (4)$$

It is worth noting that the same type of considerations (Fubini theorem) allows one to compute fairly easily the exact moments of all orders of the arithmetic average, in contrast with the unnecessary approximations which are often offered in the literature.

Formula (4) has an interesting resemblance with the Black-Scholes formula: the first term is equal to $S(t)$ times a quantity smaller than one; in the second term, the strike price k is reduced by the contribution to the average of the already observed values. However, since this second term is negative, the call price is in fact a sum of two positive terms, a property which expresses the moneyness of the option.

Obviously, in most cases, this formula does not hold since at date t the quantity $\frac{1}{T}\int_0^t Sds - k$ is likely to be non positive. To address this difficult situation in an exact approach, a solution consists (see Geman-Yor 1993) in

a) writing $S(t)$ as a time-changed squared Bessel process;

b) choosing not to compute the option price itself but rather its Laplace transform with respect to maturity, namely the function $\phi(\lambda) = \int_0^{+\infty} C(T)e^{-\lambda T}dT$.

Geman-Yor (1993) give the details of the different mathematical steps which lead to the following expression for the call price

$$C(t) = \frac{4S(t)}{\sigma^2 T}e^{-r(T-t)}C^{(\nu)}(h, q) \qquad (5)$$

where $\nu = \frac{2r}{\sigma^2} - 1; h = \frac{\sigma^2}{4}(T - t); q = \frac{\sigma^2}{4S(t)}\{kT - \int_0^t S(u)du\}$ and the Laplace transform of the quantity C^ν with respect to h is given by

$$\int_0^\infty e^{-\lambda h} C^{(\nu)}(h, q)dh = \frac{\int_0^{1/2q} dx e^{-x} x^{\frac{\mu - \nu}{2}} (1 - 2qx)^{\frac{\mu + \nu}{2} + 1}}{\lambda(\lambda - 2 - 2\nu)\Gamma(\frac{\mu - \nu}{2} - 1)} \qquad (6)$$

where Γ denotes the gamma functions and $\mu = \sqrt{2\lambda + \nu^2}$.

We can observe that when the underlying asset is a stock paying a continuous dividend y (y may also be the convenience yield of a commodity or the foreign interest rate in the case of a currency), the above results prevail exactly by replacing r by $r - y$ and ν by $\frac{2(r-y)}{\sigma^2} - 1$.

The inversion of the Laplace transform in (6) provides not only the call price but also its delta through the same methodology. Indeed, the differentiation of formula (5) with respect to S gives

$$\Delta = \frac{\partial C_t}{\partial S_t} = \frac{C_t}{S_t} - \frac{e^{-r(T-t)}}{T} \frac{1}{S(t)} \left\{ kT - \int_0^t S(u)du \right\} \frac{\partial C^\nu(h, q)}{\partial q} \qquad (7)$$

and we face an analogous problem of inversion of the Laplace transform.

Geman-Eydeland (1995) on one hand, Fu-Madan-Wang (1999) on the other hand apply different algorithms to invert the Geman-Yor formula but come up with results remarkably close (Fu-Madan-Wang use an algorithm developed by Abate and Whitt (1995); Geman-Eydeland use a method based on contour integration in the complex plane). The latter authors also provide comparisons with Monte Carlo simulations since this mathematically simple approach is very popular among practitioners and does not raise particular problems in the case of the smooth payoff of the average rate option (in contrast to barrier options).

The following table gives some numerical results (the stock is assumed to pay no dividend over the period and the date of analysis t is taken to be 0).

Interest rate r	Volatility σ	Maturity price k T	Strike value $S(0)$	Initial	G-Y	Monte-Carlo
0.05	0.5	1	2	1.9	0.195	0.191
0.05	0.5	1	2	2	0.248	0.248
0.05	0.5	1	2	2.1	0.308	0.306
0.02	0.1	1	2	2	0.058	0.056
0.0125	0.25	2	2	2	0.1772	0.1771
0.05	0.5	2	2	2	0.351	0.347

The Monte Carlo values are obtained through a sample of 50 evaluations, each evaluation being performed on 500 Monte Carlo paths.

Turning to the computation of the delta of the option, for instance for $S(0) = 2$, we know that many practitioners use an elementary finite difference method with Monte Carlo values, which means in our example a delta equal to $\frac{0.306-0.191}{0.2} = 0.575$; by doing so, a much higher error appears in the delta than in the option price itself.

On the contrary, in the Laplace transform approach and thanks to the linearity of integration and derivation, the error does not deteriorate and the delta obtained in the above example is 0.56, a number significantly different than 0.575.

To end this section, let us observe that we have addressed the so-called fixed strike Asian option. A less popular type of Asian option has a floating strike, meaning that the pay-off at maturity is expressed as $\max(A_T - S_T, 0)$. Ingersoll in his book (1987) conjectured that this case would be much simpler than the fixed-strike case. Indeed, taking the stock price as the numéraire (see Geman-El Karoui-Rochet 1995), one obtains a fairly simple partial differential equation satisfied by the Asian call option. The powerful change of numéraire technique, though still feasible, does not provide as simple a result for the fixed-strike Asian call options.

4. Barrier and Double-barrier Options

Barrier options to which a vast body of literature is currently dedicated, represent the most common type of exotic options: they were traded in over-the counter markets in the United States many years before plain vanilla options were listed (see Snyder, 1969). The pricing of "single barrier" options is not very difficult in the standard Black-Scholes-Merton framework and closed-form solutions have been available for some time. The price of a down-and-out option was already in Merton (1973) seminal paper and in 1979, Goldman-Sosin-Gatto offered explicit solutions for all types of single barrier options.

We focus in this paragraph on double-barrier options which have become very popular recently. Not only, as mentioned earlier, do they provide a less expensive hedge which may be good enough in a number of situations. But they also allow investors with a specific view on the range of a stock price without any specific anticipation on the terminal value to take a position accordingly. We will be addressing the so-called "continuously deactivating" double-barrier options (meaning that the option vanishes at any time where the underlying asset price hits the upper barrier U or the lower barrier L), as opposed to comparing the daily fixings of a stock with the numbers U and L. This is the situation which prevails in the FX markets, where double-barrier options represent a significant fraction of options written every day. The methodology described below allows to price, as a simpler case, the so-called corridor options which pay one at maturity if the underlying asset price has remained in the corridor during the lifetime of the option.

The mathematical setting Assuming that the dynamics of the underlying asset are driven under Q by the same equation (1') as before

$$\frac{dS_t}{S_t} = (r - y)dt + \sigma dW_t$$

and denoting by L the lower barrier and by U the upper barrier, we consider an option which vanishes as soon as either the upper or lower barrier is hit. The price of the call at time t is equal to

$$C(t) = E_Q[e^{-r(T-t)}\max(0, S(T) - k)1_{(\Sigma > T)}|F_t] \tag{8}$$

where $\Sigma = \inf\{t/S(t) \geq U \text{ or } S(t) \leq L\}$ is the first exit time of the process $(S(t))$ out of the interval $[L, U]$ and interest rates are supposed constant (as well as the possible dividend payment y). It is slightly easier to compute the quantity

$$D(t) = E_Q[e^{-r(T-t)}\max(0, S(T) - k)1_{(\Sigma \leq T)}|F_t]$$

Obviously, the knowledge of $D(t)$ would give $C(t)$ since the two quantities add up to the Black-Scholes price.

Again, the expression whose expectation is computed (which is a functional of the brownian motion W_t through S_T and Σ) would be simpler if the fixed maturity date was replaced by an exponential time τ independent of (W_t). This leads to compute the Laplace transform $\Psi(\lambda)$ of $D(t)$ with respect to maturity T. Geman-Yor (1996) show that

$$\Psi(\lambda) = \frac{1}{\sigma^2}\zeta(\lambda/\sigma^2)$$

where $\zeta(\lambda) = \frac{sh(\mu b)}{sh[\mu(a+b)]}g_1(e^{-a}) + \frac{sh(\mu a)}{sh[\mu(a+b)]}g_2(e^b)$ with $\alpha = \frac{1}{\sigma^2}(r - y - \frac{\sigma^2}{2})$;

$$L/S(t) = e^{-a}; \quad U/S(t) = e^b; \quad h = k/S(t); \quad \mu = \sqrt{2\lambda + \alpha^2}$$

$$g_1(e^{-a}) = \frac{h^{\alpha+1-\mu}e^{\mu a}}{\mu(\mu - \alpha)(\mu - \alpha - 1)}$$

$$g_2(e^b) = 2\left[\frac{e^{b(\alpha+1)}}{\mu^2 - (\alpha+1)^2}\right] + \frac{e^{-\mu b}h^{\alpha+1+\mu}}{\mu(\mu + \alpha)(\mu + \alpha + 1)}$$

Again, the numerical results obtained through the inversion of the Laplace transform are compared with Monte Carlo simulations. A first set of tests is performed with $t = 0, T = 1 \; year$

Parameters	$S(0) = 2$ $\sigma = 0.2$ $r = 0.02$ $k = 2,$ $L = 1.5,$ $U = 2.5$	$S(0) = 2$ $\sigma = 0.5$ $r = 0.05$ $k = 2,$ $L = 1.5,$ $U = 3$	$S(0) = 2$ $\sigma = 0.5$ $r = 0.05$ $k = 1.75,$ $L = 1,$ $U = 3$
G-Y price	0.0411	0.0178	0.07615
Monte Carlo price (st. dev $= 0.003$)	0.0425	0.0191	0.0772

where the standard deviation is computed on a sample of 200 evaluations, each evaluation being performed on 5000 Monte Carlo paths with a step size of $1/365$ year.

In order to show the nonrobustness of Monte Carlo methods when we approach maturity while the price of the underlying asset is close to one of the barriers, we take the same parameters as in the first column of the above table except that $S(0)$ is supposed to be 2.4 and the time to maturity one month. The G-Y method gives a call price equal to 0.17321 and there is no change in the accuracy nor in the computing time since the Laplace transform method is insensitive to the position of the underlying asset price with respect to the barrier. On the contrary, keeping the same step size of $1/365$ year gives a Monte Carlo standard deviation equal to 0.073 (which is clearly too high for practical purposes) and a Monte Carlo value for the call of 0.1930. By making the step four times smaller, the standard deviation is reduced to 0.008 and the price becomes 0.1739, which happens to be much closer to the G-Y price and to be lower than 0.1930 (since in the first simulations, the option may have been overpriced through some trajectories "missing" the barrier while, in reality, the underlying asset had hit it, entailing the deactivation of the option).

5. Conclusion

The methodology involving stochastic time changes and Laplace transforms has been proved to be very efficient in the valuation and hedging of the most notoriously difficult European path-dependent options, namely the Asian and double-barrier options. The results have been obtained in the classical Black-Scholes-Merton setting of a constant volatility. We can observe, however, that the introduction of a stochastic volatility in the underlying asset dynamics generally involves the use of a tree or of some numerical procedure (Monte Carlo or other). In all cases, the quasi-exact values obtained in the constant volatility case can be used as control variates to improve the accuracy of the numerical procedure.

Bibliography

Abate, J. and Whitt, W. (1995). Numerical Inversion of Laplace Transforms of Probability Distributions. *ORSA Journal of Computing*

Ane, T. and Geman, H. (2000). Order Flow, Transaction Clock and Normality of Asset Returns. *Journal of Finance*, **LV** (5), 2259–2284

Bachelier, L. (1941). Probabilités des Oscillations Maxima. *Comptes Rendus des Séances de l'Académie des Sciences*, **212**, 836–838. Erratum, **213**, 220

Black, F. and Scholes, M. (1973). The Pricing of Options and Corporate Liabilities. *Journal of Political Economy*, **81**, 637–654

Bochner, S. (1955). *Harmonic Analysis and the Theory of Probability*, University of California Press

Dufresne, D. (1990). The Distribution of a Perpetuity with Applications to Risk Theory and Pension Funding. *Scand. Act. Journal*, 39–79

Fu, M., Madan, D. and Wang, T. (1999). Pricing Continuous Time Asian Options: A Comparison of Analytical and Monte Carlo Methods. *Journal of Computational Finance*, **2**, 49–74

Geman, H., El Karoui, N. and Rochet, J.C. (1995). Changes of Numéraire, Changes of Probability Measure and Option Pricing. *Journal of Appl. Prob.*, **32**, 443–458

Geman, H. and Eydeland, A. (1995). Domino Effect: Inverting the Laplace Transform. Risk, April

Geman, H. and Yor, M. (1993). Bessel Processes, Asian Options and Perpetuities. *Mathematical Finance*, **3** (4), 349–375. **Paper [5] in this book**

Geman, H. and Yor, M. (1996). Pricing and Hedging Path-Dependent Options: A Probabilistic Approach. *Mathematical Finance*, **6** (4), 365–378

Goldman, M., Sosin, H. and Gatto, M. (1979). Path Dependent Options: Buy at the Low, Sell at the High. *Journal of Finance*, **34**, 111–127

Harrison, J.M. and Kreps, D. (1979). Martingales and Arbitrage in Multiperiod Securities Markets. *Journal of Economic Theory*, **20**, 381–408

Harrison, J.M. and Pliska, S.R. (1981). Martingales and Stochastic Integrals in the Theory of Continuous Trading. *Stoch. Proc. Appl.*, **11**, 215–260

Ingersoll, J. (1987). *Theory of Rational Decision Making*. Rowman and Littlefield

Itô, K. and McKean, H.P. (1965). *Diffusion Processes and Their Sample Paths.* Springer

Kemna, A. and Vorst, T. (1990). A Pricing Method for Options Based on Average Asset Values. *Journal of Banking and Finance*, **14**, 113–129

Kunitomo, N. and Ikeda, M. (1992). Pricing Options with Curved Boundaries. *Mathematical Finance*, **2** (4), 275–2

Kunitomo, N. and Ikeda, M. (2000). Correction: Pricing Options with Curved Boundaries. *Mathematical Finance*, **10** (4), 459

Lamperti, J. (1972). Semi-stable Markov processes, I. *Zeitschrift für Wahrsch.*, **22**, 205–225

Lévy, E. (1992). Pricing European Average Rate Currency Options. *Journal of Int. Money & Finance*, **II**, 474–491

Merton, R.C. (1973). Theory of Rational Option Pricing. *Bell. J. Econ. Manag. Sci.*, **4**, 141–183

Pitman, J. and Yor, M. (1992). The Laws of Homogeneous Functionals of Brownian Motion and Related Processes. Preprint, University of California at Berkeley.

Revuz, D. and Yor, M. (1998). *Continuous Martingales and Brownian motion.* 3rd edition, Springer

Rogers, L.C.G. and Shi, Z. (1995). The Value of an Asian Option. *Journal of Appl. Prob.,* **32**, 1077–1088

Ross, S. (1976). The Arbitrage Theory of Capital Asset Pricing. *Journal of Economic Theory,* **13** (3), 341–360

Yor, M. (1992). Sur Certaines Fonctionnelles Exponentielles du Mouvement Brownien Réel. *Journal of Appl. Prob.,* **29**, 202–208. **Paper [1] in this book**

On Certain Exponential Functionals
of Real-Valued Brownian Motion

J. Appl. Prob. **29** (1992), 202–208

Abstract. Dufresne [1] recently showed that the integral of the exponential of Brownian motion with negative drift is distributed as the reciprocal of a gamma variable. In this paper, it is shown that this result is another formulation of the distribution of last exit times for transient Bessel processes. A bivariate distribution of such integrals of exponentials is obtained explicitly.

1. Introduction

1.1. Let $(B_t, t \geq 0)$ be a Brownian motion, starting at 0. In studies of financial mathematics carried out over the last decade, the processes

$$\exp(aB_t - bt), \quad t \geq 0, \text{ with } a, b \in \mathbb{R}, \tag{1.a}$$

play an important role, for example, in the Black–Scholes formula.

In May 1990, at the time of the Itô Colloquium in Paris, M. Emery told me about the work of Dufresne [1], a Canadian actuary, who had just shown, by somewhat complicated means, that the variable

$$\int_0^\infty ds \exp(aB_s - bs), \quad \text{where } a \neq 0, b > 0, \tag{1.b}$$

is distributed as the reciprocal of a gamma variable, up to a multiplicative constant.

The main purpose of this note is to show that this result is another formulation of the following result due to Getoor [2]: if $(\hat{R}_t, t \geq 0)$ is a Bessel process, starting at 0, of dimension $\hat{\delta} = 2(1 + \nu)$, with $\nu > 0$, then:

$$L_1(\hat{R}) \stackrel{\text{dist.}}{=} \frac{1}{2Z_\nu} \tag{1.c}$$

where $L_1(\hat{R}) \equiv \sup\{t > 0 : \hat{R}_t = 1\}$, and Z_ν is a gamma variable with index ν; in other words, it satisfies

$$P(Z_\nu \in dt) = \frac{t^{\nu-1} e^{-t}}{\Gamma(\nu)} dt \quad (t \geq 0). \tag{1.d}$$

This connection between the two results is partially explained by the fact that if $(R_t, t \geq 0)$ is a Bessel process, starting at 1, of dimension $\delta = 2(1+\mu)$ (here, μ is no longer necessarily positive), then:

$$R_t = \exp(\beta_u + \mu u)|_{u=H_t}, \quad \text{with } H_t = \int_0^t \frac{ds}{R_s^2}, \tag{1.e}$$

and β is a real-valued Brownian motion starting from 0.

The equivalence of the results of Dufresne and Getoor will be demonstrated in Section 2, below.

1.2. It then seemed interesting to study the joint law of the variables appearing in (1.b) when a and b vary; however, we did not arrive at truly explicit results, except in relation to the two-dimensional vector

$$(X_+; X_-) \equiv \left(\int_0^\infty ds \exp(2B_s - s); \int_0^\infty ds \exp(-2B_s - s) \right), \tag{1.f}$$

whose distribution is studied in Section 3.

2. Two Realizations of the Reciprocal of a Gamma Variable

2.1. We first give a precise statement of Dufresne's result in Theorem 1 and Corollary 1.

Theorem 1. *For all $\nu > 0$, we have*

$$\int_0^\infty dt \exp\left(B_t - \frac{\nu t}{2} \right) \stackrel{\text{dist.}}{=} \frac{2}{Z_\nu}, \tag{2.a}$$

where Z_ν is a gamma variable with index ν (see formula (1.d)).

Using the scaling and symmetry properties of the Brownian motion, we deduce the following result from Theorem 1.

Corollary 1. *For all $a \in \mathbb{R}$, $a \neq 0$, and $b > 0$, we have:*

$$\int_0^\infty ds \exp\left(aB_s - bs \right) \stackrel{\text{dist.}}{=} 2/a^2 Z_{2b/a^2}. \tag{2.b}$$

The result (2.b) can also be presented in terms of Bessel processes.

Corollary 2. *Let $(R_t, t \geq 0)$ be a Bessel process, starting at 1, of dimension $\delta = 2(1+\mu)$, with $\mu > 0$.*
Then for all $\alpha > 0$, we have:

$$\int_0^\infty \frac{ds}{R_s^{2+\alpha}} \stackrel{\text{dist.}}{=} 2/\alpha^2 Z_{2\mu/\alpha}. \tag{2.c}$$

Proof. According to (1.e), we have

$$\int_0^\infty \frac{ds}{R_s^{2+\alpha}} = \int_0^\infty du \exp -\alpha(\beta_u + \mu u),$$

and it is therefore sufficient to apply Corollary 1 with $a = -\alpha$, and $b = \alpha\mu$. $\square$

2.2. We now prove Theorem 1, using the result (1.c) due to Getoor.

Let $\nu > 0$. Let us apply Itô's formula to $\exp(B_t - \nu t)$, $t \geq 0$. We obtain:

$$\exp(B_t - \nu t) = 1 + \int_0^t \exp(B_s - \nu s)dB_s$$
$$+ \left(\frac{1}{2} - \nu\right) \int_0^t ds \exp(B_s - \nu s). \quad (2.d)$$

Let us now set:

$$A_t = \int_0^t ds \exp 2(B_s - \nu s),$$

and define the process $(\rho_u, u < A_\infty)$ via

$$\exp(B_t - \nu t) = \rho_{A_t}. \quad (2.e)$$

From (2.d), we deduce the identity

$$\rho_u = 1 + \gamma_u + \left(\frac{1}{2} - \nu\right) \int_0^u \frac{ds}{\rho_s} \quad (2.f)$$

where $(\gamma_u, u < A_\infty)$ is a real-valued Brownian motion; in other words, ρ is a Bessel process, starting at 1, of dimension $\delta = 2(1 - \nu)$. We now note that, according to (2.e), as $u \to A_\infty$, $\rho_u \to 0$. It is then possible to extend $(\rho_u, u < A_\infty)$ by continuity to time A_∞, which then appears as: $A_\infty = T_0(\rho) = \inf\{u : \rho_u = 0\}$. Now, according to classical results on time reversal (cf. Williams [6], Sharpe [5]), we have:

$$(\rho_{T_0 - u}; u \leq T_0(\rho)) \overset{\text{dist.}}{=} (\hat{\rho}_u; u \leq L_1(\hat{\rho})), \quad (2.g)$$

where $(\hat{\rho}_u, u \geq 0)$ is a Bessel process, starting from 0, of dimension $\hat{\delta} = 2(1+\nu)$ and $L_1(\hat{\rho}) = \sup\{t > 0 : \hat{\rho}_t = 1\}$. Thus, we have:

$$A_\infty \overset{\text{dist.}}{=} T_0(\rho) \overset{\text{dist.}}{=} L_1(\hat{\rho}),$$

and the desired result is then obtained by virtue of (1.c) (more precisely, we have proved here the result (2.b) for $a = 2$, and $b = 2\nu$, and the result (2.a) follows by scaling).

2.3. For completeness, we now give an elementary proof of Getoor's result [2]:

$$L_1(\hat{R}) \stackrel{\text{dist.}}{=} \frac{1}{2Z_\nu} \tag{1.c}$$

where $\hat{R}$ denotes a Bessel process, starting at 0, of dimension $\hat{\delta} = 2(1+\nu)$. We begin by proving the intermediate identity in distribution

$$L_1(\hat{R}) \stackrel{\text{dist.}}{=} \left(\inf_{u \geq 1} \hat{R}_u \right)^{-2}. \tag{2.h}$$

In fact, for all $t > 0$, we have:

$$(L_1(\hat{R}) > t) = \left(\inf_{u \geq t} \hat{R}_u < 1 \right)$$

$$\stackrel{\text{dist.}}{=} \left(\sqrt{t} \inf_{u \geq 1} \hat{R}_u < 1 \right), \text{by scaling}$$

$$\stackrel{\text{dist.}}{=} \left(\left(\inf_{u \geq 1} \hat{R}_u \right)^{-2} > t \right),$$

which implies (2.h).

It now remains to identify the distribution of $\inf_{u \geq 1} \hat{R}_u$ which will follow easily from the *maximal equality*, given in the following lemma.

Lemma 1. *Let $(M_t, t \geq 0)$ be a local continuous martingale with values in $\mathbb{R}_+$, which tends to 0 as $t \to \infty$. Then*

$$\sup_{t \geq 0} M_t \stackrel{\text{dist.}}{=} \frac{M_0}{U}, \tag{2.i}$$

where U is a uniformly distributed variable on $(0, 1)$, independent of M_0.

The proof of Lemma 1 follows immediately from the stopping theorem applied to $T_a = \inf\{t : M_t \geq a\}$, which gives:

$$P\left\{ \sup_{t \geq 0} M_t \geq a \Big| \mathcal{F}_0 \right\} = (M_0/a) \wedge 1,$$

whence we deduce (2.i).

Let us now return to (2.h) and apply Lemma 1 to $M_t = 1/(\hat{R}_t)^{\hat{\delta}-2}$, $t \geq 1$. Then, following (2.i), we have:

$$\inf_{t \geq 1} \hat{R}_t \stackrel{\text{dist.}}{=} \hat{R}_1 U^{1/\hat{\delta}-2},$$

where U is a uniform variable on $(0, 1)$, independent of $\hat{R}_1$. Following (2.h), we therefore have

$$\frac{1}{L_1(\hat{R})} \overset{\text{dist.}}{=} (\hat{R}_1)^2 U^{1/\nu}.$$

We then obtain the result (1.c) by applying (for example!) the algebraic formula on beta–gamma variables, in other words:

$$Z_a \overset{\text{dist.}}{=} Z_{a+b} Z_{a,b},$$

where, on the right-hand side, the variable Z_{a+b} is independent of $Z_{a,b}$, a beta variable with parameters (a, b). Then, we have:

$$(\hat{R}_1)^2 \overset{\text{dist.}}{=} 2Z_{\hat{\delta}/2},$$

and it is sufficient to take a and b such that $a + b = \hat{\delta}/2$, and $b = 1$, whence: $a = \nu$.

3.   The Distribution of the Vector (X_+, X_-)

The key to the results of this section is the following lemma, in the same vein as Lemma 1.

Lemma 2. *Let $(M_t, t \geq 0)$ be a continuous local martingale, with values in $\mathbb{R}_+$, starting at $a > 0$, which tends to 0 as $t \to \infty$. Then we have:*

$$\langle M \rangle_\infty \overset{\text{dist.}}{=} T_a \equiv \inf\{t : B_t = a\},$$

where B denotes a real-valued Brownian motion, starting at 0.

This lemma follows directly from the Dubins–Schwarz representation of M as a Brownian motion, with a time change, by means of $(\langle M \rangle_t, t \geq 0)$. One consequence of Lemma 2 is the following result.

Corollary 3. *Let $(B_t, t \geq 0)$ be a real-valued Brownian motion, starting at 0, and*

$$T_a \equiv \inf\{t : B_t = a\} \quad (a \geq 0).$$

Then, for any n-tuplet of reals $(\alpha_1, \ldots, \alpha_n)$, and any n-tuplet of positive reals $(a_1, \ldots, a_n)$, we have:

$$\sum_{i,j=1}^{n} a_i a_j \alpha_i \alpha_j \int_0^\infty ds \exp\left((\alpha_i + \alpha_j) B_s - \frac{\alpha_i^2 + \alpha_j^2}{2} s\right) \overset{\text{dist.}}{=} T_{(\Sigma_{i=1}^n a_i)}, \qquad (3.a)$$

Proof. We apply Lemma 2 to:

$$M_t = \sum_{i=1}^{n} a_i \exp\left(\alpha_i B_t - \frac{\alpha_i^2 t}{2}\right), \quad t \geq 0.$$

$\square$

Unfortunately, the identity in distribution (3.a) does not allow us to calculate, at least not in a direct way, the joint law of the functionals

$$\left(\int_0^\infty ds \exp\left((\alpha_i + \alpha_j)B_s - \frac{\alpha_i^2 + \alpha_j^2}{2}s\right); 1 \leq i, j \leq n\right).$$

However, in the case $n = 2$, $\alpha_1 = 1$, $\alpha_2 = -1$, we have the following proposition.

Proposition. *Let $(B_t, t \geq 0)$ be a Brownian motion, starting at 0; set:*

$$X_+ = \int_0^\infty ds \exp(2B_s - s) \text{ and } X_- = \int_0^\infty ds \exp(-2B_s - s).$$

Then, for all $a, b \geq 0$, we have:

$$E\left[\exp{-\frac{1}{2}(a^2 X_+ + b^2 X_-)}\right] = \exp{-(a + b + ab)}, \tag{3.b}$$

which can be rewritten in the form:

$$(X_+, X_-) \overset{\text{dist.}}{=} \left(T_+, \frac{1}{T_+} + \frac{T_-}{T_+^2} + V\right), \tag{3.c}$$

where on the right-hand side of (3.c) the variables T_+, T_- and V are independent and T_+ and T_- are distributed as $T_1 \equiv \inf\{t : B_t = 1\}$, and the Laplace transform of V is given by

$$E\left[\exp\left(-\frac{\lambda^2 V}{2}\right)\right] = (1 + \lambda)e^{-\lambda} \quad (\lambda \geq 0). \tag{3.d}$$

Proof. Following (3.a), we have:

$$a^2 X_+ + b^2 X_- - 2ab \overset{\text{dist.}}{=} T_{(a+b)},$$

which implies the formula (3.b); the identity in distribution (3.c) then follows from the fact that by virtue of (3.b) we have:

$$E\left[\exp\left(-\frac{b^2 X_-}{2}\right)\Big| X_+ = t\right] = \exp\left(-\frac{b^2}{2t}\right)(1 + b)\exp\left(-b\left(1 + \frac{1}{t}\right)\right).$$

$\square$

Remark. It is not difficult to show, using formula (3.d), by taking the derivatives on both sides of the following identity at $a = 1$:

$$e^{-\lambda a} = \int_0^\infty \frac{dt}{\sqrt{2\pi t^3}} a \exp(-(a^2/2t) - \lambda^2 t/2)$$

together with Getoor's result (1.c), that we have:

$$V \overset{\text{dist.}}{=} L_1(R^{(5)}) \overset{\text{dist.}}{=} \frac{1}{2Z_{3/2}} , \qquad (3.e)$$

where $R^{(5)}$ denotes a Bessel process of dimension five, starting at 0.

4.   Comments

4.1. In this paper we have seen three different representations of the reciprocal of a gamma variable:

- as an exponential functional of Brownian motion (Theorem 1; this is Dufresne's result [1]),
- as the last passage time through 1 for a transient Bessel process (result due to Getoor [2]),
- as the integral of a negative power of a Bessel process (this is Corollary 2).

The relationships between the second and third interpretations had already been brought to light in the study of the (geometrical) inverse of Brownian motion in $\mathbb{R}^n$ [7].

4.2. In [3] (see also [4], Exercise (4.16), p. 298), one can find a general formula which can be used to give an explicit expression for the distribution of the last passage time through a point for a transient diffusion with values in $\mathbb{R}_+$; Getoor's result (1.c) then appears as a special case of this general formula.

4.3. It may be of interest to note here that, with the notation of the proof of Theorem 1 in Section 2.2, conditionally on $T_0(\rho) = t$, the process $(\rho_u, u \le t)$ is a Bessel bridge of dimension $\delta = 2(1 - \nu)$ on the interval $[0, t]$; similarly, $(\hat{\rho}_u, u \le t)$ is a Bessel bridge of dimension $\hat{\delta} = 2(1 + \nu)$ on the interval $[0, t]$. A detailed discussion can be found in [7]; see also, more generally, Revuz and Yor [4], Exercise (1.16), p. 378.

4.4. Corollary 3 has served as the starting point for a detailed study of stable multivariate variables with index $\frac{1}{2}$ or of certain multivariate Cauchy variables [8].

Acknowledgments

I am very grateful to D. Dufresne for providing a preprint of his work, to M. Emery for making me aware of Dufresne's result and also to A. Joffe, W. Kendall and G. Letac for a number of conversations on this subject. W. Kendall and G. Letac also have proofs which differ significantly from Dufresne's.

References

1. Dufresne, D. (1990). The distribution of a perpetuity, with applications to risk theory and pension funding. *Scand. Actuarial J.*, 39–79
2. Getoor, R.K. (1979). The Brownian escape process. *Ann. Probab.*, **7**, 864–867
3. Pitman, J.W. and Yor, M. (1981). Bessel processes and infinitely divisible laws. In 'Stochastic Integrals' (Lecture Notes in Mathematics, vol. 851) ed. D. Williams, Springer, Berlin, 285–370
4. Revuz, D. and Yor, M. (1991). Continuous Martingales and Brownian Motion, Springer, Berlin
5. Sharpe, M. (1980). Some transformations of diffusions by time reversal. *Ann. Probab.*, 1157–1162
6. Williams, D. (1974). Path decomposition and continuity of local time for one-dimensional diffusions. *Proc. Lond. Math. Soc.*, **28**, 738–768
7. Yor, M. (1984). A propos de l'inverse du mouvement brownien dans $\mathbb{R}^n$ ($n \geq 3$). *Ann. Inst. Henri Poincaré, Probab. Stat.*, **21**, 27–38
8. Yor, M. (1990). Sur les décompositions affines d'une variable stable d'indice $\frac{1}{2}$. [*See point* **e**) *of the Postscript*]

Postscript #1

a) In May 2000, D. Siegmund told me about the existence of the two following papers, which relate exponential functionals to changes of drift for Brownian motion (with drift):

   M. Pollak, D. Siegmund (1985): "*A diffusion process and its applications to detecting a change in the drift of Brownian motion*". Biometrika 72, p. 267–280.

   H. Rubin, K. S. Song (1995): "*Exact computation of the asymptotic efficiency of maximum likelihood estimators for a discontinuous signal in a Gaussian white noise*". Ann. Stat. 23, p. 732–739.

   In particular, Theorem 1 in the present paper is closely related to Proposition 3 in the Pollak-Siegmund paper.

b) A number of results in the present monograph hinge upon the following representation formula due to Lamperti (1972):

$$\exp(\xi_t) = X_{\left(\int_0^t ds \exp(\xi_s)\right)} \tag{1}$$

where $(\xi_t, t \geq 0)$ is a general Lévy process and $(X_u, u \geq 0)$ a Markov process with self-similarity property of order 1.

In this monograph, this important result of Lamperti is only cited in the article [8], which deals with Lévy processes. When $(\xi_t, t \geq 0)$ is a Brownian motion with drift ν $(\in \mathbb{R})$, then $(X_u, u \geq 0)$ is a squared Bessel process of index ν, and this result is part of the common knowledge concerning one-dimensional diffusions.....

c) When $\nu = -\mu$, $\mu > 0$, it may appear as a "coïncidence" that the law featured in Theorem 1 is that of the reciprocal of γ_μ (γ_μ denotes a $gamma(\mu)$ variable), whereas the law of a squared Bessel process of index μ, at time 1, is that of $2\gamma_{\mu+1}$. In fact, the two papers:

J. Bertoin, M. Caballero: "*Entrance from 0^+ for increasing semi-stable Markov process*". (Submitted 2000)

J. Bertoin, M. Yor: "*The entrance laws of self-similar Markov processes and exponential functionals of Lévy processes*". (Submitted January 2001)

present a formula which extends the validity of this "coïncidence" to a large class of Lévy processes $(\xi_t, t \geq 0)$ represented as in (1).

d) About section 3: In this section, the problem of finding out the laws of multidimensional perpetuities is posed. Some variant of this problem already appeared in the Rubin-Song paper quoted above in a). So far, this problem is mostly unsolved; some minor progress is made in:

P. Carmona, F. Petit, M. Yor (2001): "*Exponential functionals of Lévy processes*". In O. Barndorff-Nielsen, T. Mikosch, S. Resnick (eds.): "*Lévy processes : theory and applications*". Birkhäuser, p. 41–55.

e) The reference 8. (above) finally appeared, in a much more developed form than the original draft as:

C. Donati-Martin, R. Ghomrasni, M. Yor, Affine random equations and the stable (1/2) distribution.

Studia Scient. Math. Hung **36** (Oct. 2000), 387–405.

On Some Exponential Functionals
of Brownian Motion

Adv. Appl. Prob. **24** (1992), 509–531

Abstract. In this paper, distributional questions which arise in certain mathematical finance models are studied: the distribution of the integral over a fixed time interval $[0, T]$ of the exponential of Brownian motion with drift is computed explicitly, with the help of computations previously made by the author for Bessel processes. The moments of this integral are obtained independently and take a particularly simple form. A subordination result involving this integral and previously obtained by Bougerol is recovered and related to an important identity for Bessel functions. When the fixed time T is replaced by an independent exponential time, the distribution of the integral is shown to be related to last-exit time distributions and the fixed time case is recovered by inverting Laplace transforms.

1. Introduction

1.1. For the last four or five years, a number of applied probabilists, working in the domain of mathematical finance, and more precisely on path-dependent options, the so called 'Asian options', a particular case of which are average-value options, have been interested in the distribution of

$$\int_0^t ds \, \exp(aB_s + bs), \tag{1.a}$$

for a fixed time t, and reals a and b, where $(B_s, s \geqq 0)$ denotes a real-valued Brownian motion starting from 0.

Thanks to the scaling properties of Brownian motion, it suffices to find this distribution for $a = 2$ for instance. We shall now write

$$A_t^{(\nu)} = \int_0^t ds \, \exp 2(B_s + \nu s), \tag{1.b}$$

and discuss the distribution of $A_t^{(\nu)}$. When $\nu = 0$, we simply write A_t. At this point, it may be worth emphasizing that some quantities of interest in mathematical finance are the moments of $A_t^{(\nu)}$, and perhaps more importantly, the function

$$C(t, k) \equiv E((A_t^{(\nu)} - k)^+). \tag{$*$}$$

We come back to this point in detail in Section 1.7 of this introduction.

Hence a fairly thorough knowledge of the law of $A_t^{(\nu)}$ is needed in order to get, as much as possible, an explicit expression of $C(t, k)$.

1.2. We now remark that, thanks to Girsanov's theorem, we have, for any Borel function $f : \mathbb{R}_+ \longrightarrow \mathbb{R}_+$:

$$E[f(A_t^{(\nu)})] = E\left[f(A_t)\exp\left(\nu B_t - \frac{1}{2}\nu^2 t\right)\right] \tag{1.c}$$

so that the distribution of $A_t^{(\nu)}$ may be obtained once the joint distribution of $(A_t, \exp(B_t))$ is known. This may be done in terms of the semigroup of the hyperbolic Brownian motion on Poincaré's half-plane (see Bougerol [2]).

In order to be complete and, at the same time, not to confuse the reader with different approaches, the connections with hyperbolic Brownian motion are presented concisely in Section 7 of the present paper.

However, the main approach chosen in this paper, for the study of the law of A_t, and more generally $A_t^{(\nu)}$, is by relating $(\exp(B_t + \nu t), t \geq 0)$ and the Bessel process $(\rho_u^{(\nu)}, u \geq 0)$ with index ν. Some important facts concerning Bessel processes and related computations are collected in Section 2.

1.3. An interesting by-product of this computation is the following striking identity in law obtained by Bougerol [2]:

$$\text{for any fixed } t > 0, \quad \sinh(B_t) \stackrel{(\text{law})}{=} \gamma_{A_t}, \tag{1.d}$$

where $(\gamma_u, u \geq 0)$ is a one-dimensional Brownian motion starting from 0, and independent from B.

Routine computations show that (1.d) is equivalent to the relation

for $u \in \mathbb{R}$,

$$E\left[\frac{1}{\sqrt{A_t}}\exp\left(-\frac{u^2}{2A_t}\right)\right] = \frac{1}{\sqrt{(1+u^2)t}}\exp{-\frac{1}{2t}(\text{Arg sh } u)^2} \tag{1.e}$$

an identity which characterizes the law of A_t.

In Section 3, we show that both sides of (1.e) admit the same Laplace transform (in t) thereby providing (1.e), hence (1.d). The key argument which we use to prove this Laplace transform identity in law is:

$$\exp(B_t) = \rho_{A_t} \quad (t \geq 0) \tag{1.f}$$

where ρ is a two-dimensional Bessel process starting from 1.

It finally turns out that Bougerol's identity in law may be understood as a probabilistic interpretation of the following special case of the Lipschitz–Hankel formulae (see Watson [13], p. 386, for example): for $a \geq 1$, and $\nu > 0$, we have

$$\nu \int_0^\infty \frac{dt}{t}\exp(-at)I_\nu(t) = \frac{1}{(a + \sqrt{a^2 - 1})^\nu}.$$

1.4. In Section 4, we show, using elementary arguments independent of the previous computations, that there exists an explicit sequence of polynomials $(P_n, n \in \mathbb{N})$ such that:

$$E\left[\left(\int_0^t ds \exp(2B_s)\right)^n\right] = \frac{1}{4^n} E[P_n(\exp 2B_t)]. \qquad (1.g)$$

In particular, this formula allows us to compute the moments of A_t, and it is easily shown that formula (1.g) agrees with Bougerol's formula (1.d).

However, we also show, in the same paragraph, that the knowledge of the moments of A_t does not *a priori* determine the law of A_t, since Carleman's criterion does not apply. Hence, our derivation of formula (1.g) in Section 4, is, *a priori*, strictly weaker than Bougerol's result (1.d).

1.5. In Section 5, we compute the law of $A_{S_\theta}^{(\nu)}$, where S_θ is an exponential random variable, with parameter $(\frac{1}{2}\theta^2)$, which is independent of B. For simplicity, we now discuss the case $\nu = 0$.

We first recall that Williams [14] showed that, given $B_{S_\theta} = x > 0$, the process $(B_u, u \leq S_\theta)$ is distributed as $(B_u + \theta u; u \leq L_x^\theta)$, where $L_x^\theta \equiv \sup\{u : B_u + \theta u = x\}$; changing B into $-B$, it immediately follows that, given $B_{S_\theta} = x < 0$, the process $(B_u, u \leq S_\theta)$ is distributed as $(B_u - \theta u; u \leq L_x^{-\theta})$, where $L_x^{-\theta} \equiv \sup\{u : B_u - \theta u = x\}$. Now, the time change identity (1.f) may also be extended to

$$\exp(B_t + \theta t) = \rho_{A_t^{(\theta)}}^{(\theta)}, \qquad (1.f)_\theta$$

where $(\rho_u^{(\theta)}; u \geq 0)$ denotes a Bessel process with index θ (or dimension $d_\theta \equiv 2(\theta + 1)$) starting from 1.

As a consequence of both (1.f) and (1.f)$_\theta$, we obtain that, given $B_{S_\theta} = x > 0$, we have:

$$A_{S_\theta} \stackrel{\text{(law)}}{=} \Lambda_{e^x}^{(\theta)}, \qquad (1.h)$$

where $\Lambda_a^{(\theta)} \equiv \sup\{u > 0 : \rho_u^{(\theta)} = a\}$.

Moreover, as is well known in the general context of transient diffusions (see Pitman and Yor [9], for example), the law of $\Lambda_a^{(\theta)}$ is intimately linked with the semigroup of $\rho^{(\theta)}$, which is known explicitly. Hence, an explicit formula for the law of A_{S_θ} is easily deduced from formula (1.h).

1.6. In Section 6, we invert the Laplace transforms obtained in Section 5 to derive the law of A_t, given B_t, from computations done in [15], relative to the two-dimensional Bessel process $(\rho_u, u \geq 0)$.

1.7. We now give detailed explanations as to why the knowledge of the distribution of $A(t)$ is most interesting when dealing with average-value (financial) options (in what follows, we shall write AV-options for short).

The following presentation of such options is taken from Kemna and Vorst [7]: consider a perfect security market which is open continuously, offers a constant riskless interest rate r to all borrowers and lenders, and in which no transaction costs and/or takes are incurred. It is further assumed that the underlying asset on which the option is based is equal to a stock with price $S(t)$, which satisfies the linear stochastic differential equation

$$dS_t = \alpha S_t dt + \sigma S_t dB_t, \qquad S_0 = 1, \tag{1.i}$$

where α and σ are non-negative constants. Hence, we deduce from (1.i) that

$$S_t = \exp\left(\sigma B_t + \alpha t - \frac{1}{2}\sigma^2 t\right). \tag{1.j}$$

We then introduce the process

$$\mathcal{A}(t) = \frac{1}{T - T_0} \int_{T_0}^{t} du S_u, \quad T_0 \leq t \leq T, \tag{1.k}$$

where T is the maturity date, and $[T_0, T]$ is the final time interval over which the average value of the stock is calculated. The payoff on the AV-option can be expressed as $(\mathcal{A}(T) - K)^+$, where K is the exercise price of the option.

Kemna and Vorst ([7], p. 121) then show that the value of the AV-option at time $t \in [T_0, T]$ is given by

$$\tilde{C}(S(t), \mathcal{A}(t), t) = \exp(-r(T - t))E[(\mathcal{A}(T) - K)^+ | \mathcal{F}_t] \tag{1.l}$$

where $\mathcal{F}_t$ is the σ-field generated by the past of S, up to time t.

Thanks to the independence of the increments of B, it is immediate that the quantity (1.l) may be expressed in terms of the function C defined above in $(*)$, but now k is an explicit expression depending on $\mathcal{A}(t)$.

Because of the current lack of knowledge about the law of A_t, and more generally of $A_t^{(\nu)}$, different authors working on this subject have proceeded to various approximations (see, for example, Bouaziz et al. [1], Vorst [12]) and Monte Carlo simulations (Kemna and Vorst [7], who even made the very strange claim that 'it is impossible to derive an explicit analytic expression for an AV-option'). In a future publication, we hope to develop fully the computation of $C(t, k)$, defined in $(*)$, with the help of our knowledge of the law of $A_t^{(\nu)}$.

The main advantage of the AV-options is that they depend on the entire past history of the market, and hence, at least heuristically, reduce the risk of price manipulations of the underlying asset at the maturity date. This may explain why, more generally, path-dependent options seem to be widely used on some (Asian, but also Western) marketplaces, in particular to protect corporations against potentially hostile takeovers (see Bouaziz et al. [1] and Kemna and Vorst [7] for actual examples).

A second advantage of AV-options is that the price of an AV-option is always strictly less than the price of the corresponding standard European option; in mathematical terms, this is expressed by the inequality

$$E[(\mathcal{A}(T) - K)^+] \leqq E[(S_T - K)^+] \tag{1.m}$$

which we now show to be a simple consequence of Jensen's inequality. Indeed, we have, trivially,

$$(\mathcal{A}(T) - K)^+ \leqq \frac{1}{T - T_0} \int_{T_0}^{T} du(S_u - K)^+$$

and, since $(S_u, u \geqq 0)$ is a submartingale, the inequality

$$(S_u - K)^+ \leqq E[(S_T - K)^+ | \mathcal{F}_u] \qquad (u \leqq T). \tag{1.n}$$

holds. The inequality (1.m) follows by taking expectations of both sides of (1.n), and then averaging over $[T_0, T]$.

1.8. To conclude this introduction, we should like to sum up the contents of this paper as follows: the law of A, and more generally $A^{(\nu)}$, taken at a fixed time or at an independent exponential time, is obtained, and characterized in several ways, whilst the actual computation of $C(t, k)$ (see $(*)$) is being postponed.

Moreover, the moments of $A_t^{(\nu)}$ are obtained explicitly (see Corollary 2 of Theorem 1).

2. Some Preliminaries on Bessel Processes

The information about Bessel processes which is presented in this section may be found in Yor [15] and Pitman and Yor [9].

2.1. The key fact in the present paper is the relationship

$$\exp(B_t + \nu t) = \rho_{A_t^{(\nu)}}^{(\nu)}, \quad t \geqq 0, \tag{2.a}$$

where $(\rho_u^{(\nu)}, u \geqq 0)$ is a Bessel process with index ν, that is, an $\mathbb{R}_+$-valued diffusion with infinitesimal generator $\mathcal{L}^\nu$ given by

$$\mathcal{L}^\nu f(x) = \frac{1}{2} f''(x) + \frac{2\nu + 1}{2x} f'(x), \quad f \in C_b^2(\mathbb{R}_+^*).$$

2.2. For any $\nu \in \mathbb{R}$, and $a \geq 0$, we denote by P_a^ν the law on $C(\mathbb{R}_+, \mathbb{R}_+)$ of $\rho^{(\nu)}$, when starting from a. We write $(R_t, t \geq 0)$ for the canonical process on $C(\mathbb{R}_+, \mathbb{R}_+)$ and we denote by $\{\mathcal{R}_t = \sigma(R_s, s \leq t), t \geq 0\}$ the canonical filtration.

Using Girsanov's theorem, it is easily shown that, for $\nu \geq 0$, the mutual absolute continuity relation holds, for $a > 0$:

$$P_{a|\mathcal{R}_t}^\nu = \left(\frac{R_t}{a}\right)^\nu \exp\left(-\frac{\nu^2}{2}\int_0^t \frac{ds}{R_s^2}\right) \cdot P_{a|\mathcal{R}_t}^0. \qquad (2.\text{b})$$

In the case $\nu < 0$, the Bessel process $(\rho_u^{(\nu)}, u \geq 0)$ reaches 0, hence the law $P_{a|\mathcal{R}_t}^\nu$ cannot be equivalent to $P_{a|\mathcal{R}_t}^0$; nonetheless we have, for $\nu < 0$, and $a > 0$,

$$P_{a|\mathcal{R}_t \cap (t < T_0)}^\nu = \left(\frac{R_t}{a}\right)^\nu \exp\left(-\frac{\nu^2}{2}\int_0^t \frac{ds}{R_s^2}\right) \cdot P_{a|\mathcal{R}_t}^0, \qquad (2.\text{b}')$$

with $T_0 \equiv \inf\{u > 0 : R_u = 0\}$. (Note that the formula $(2.\text{b}')$ is also valid for $\nu \geq 0$, since then $P_a^\nu(T_0 = \infty) = 1$.) Comparing formulae $(2.\text{b})$ and $(2.\text{b}')$, we also obtain: for $\nu \geq 0$, and $a > 0$,

$$P_{a|\mathcal{R}_t \cap (t < T_0)}^{-\nu} = \left(\frac{a}{R_t}\right)^{2\nu} \cdot P_{a|\mathcal{R}_t}^\nu. \qquad (2.\text{c})$$

2.3. In what follows, we shall also need the explicit form of the density of the Bessel semigroup $P_t^\nu(a, dr) = p_t^\nu(a, r)dr$, for $\nu \geq 0$; these densities are known to be

$$p_t^\nu(a, r) = \left(\frac{r}{a}\right)^\nu \frac{r}{t} \exp\left(-\frac{1}{2t}(a^2 + r^2)\right) I_\nu\left(\frac{ar}{t}\right), \qquad (2.\text{d})$$

for $a > 0, r \geq 0$, and $t > 0$ (see, for example, Revuz and Yor [10], p. 415).

Comparing formulae $(2.\text{b})$ and $(2.\text{d})$, we now obtain the important relationship

$$E_a^0\left[\exp\left(-\frac{\nu^2}{2}\int_0^t \frac{ds}{R_s^2}\right) | R_t = r\right] = \left(\frac{I_{|\nu|}}{I_0}\right)\left(\frac{ar}{t}\right). \qquad (2.\text{e})$$

This formula $(2.\text{e})$ gives a probabilistic interpretation for the Hartman-Watson probability measure on $\mathbb{R}_+$, which we denote as $\eta_r(du)$, for $r > 0$, and which is defined and characterized by

$$\left(\frac{I_{|\nu|}}{I_0}\right)(r) = \int_0^\infty \exp\left(-\frac{\nu^2 u}{2}\right) \eta_r(du), \quad \text{for } \nu \in \mathbb{R}. \qquad (2.\text{e}')$$

For more probabilistic interpretations of the (originally) purely analytic work of Hartman and Watson [4], we refer the reader to Yor [15], and Pitman and Yor [9].

2.4. Finally, we shall make some use in what follows of the formulae (2.f) and (2.g) below involving Bessel functions:

(i) Formula (2.f) below is a particular case of the more general Lipschitz–Hankel formulae (see, for example, Watson [13], p. 386, formula (7)): for $\nu \geq 0$, and $a \geq 1$,

$$\nu \int_0^\infty \frac{dt}{t} \exp(-at) I_\nu(t) = \frac{1}{(a + \sqrt{a^2 - 1})^\nu}. \tag{2.f}$$

This formula is also found in the literature in the slightly different, but equivalent form: for $\nu \geq 0$, and $\theta \geq 0$,

$$\nu \int_0^\infty \frac{dt}{t} \exp(-(\theta + 1)t) I_\nu(t) = \frac{2^\nu}{(\sqrt{\theta + 2} + \sqrt{\theta})^{2\nu}}. \tag{2.f'}$$

(ii) We also need to recall the classical integral representation for the Bessel function K_ν: for any $\nu \in \mathbb{R}$,

$$K_\nu(z) = \frac{1}{2} \left(\frac{z}{2}\right)^\nu \int_0^\infty \frac{dt}{t^{\nu+1}} \exp - \left(t + \frac{z^2}{2t}\right) \tag{2.g}$$

(see, for example, Watson [13], p. 183, formula (15)). We shall also use the well-known formulae

$$K_{\frac{1}{2}}(z) = K_{-\frac{1}{2}}(z) = \left(\frac{\pi}{2z}\right)^{\frac{1}{2}} e^{-z}. \tag{2.g'}$$

3.   A Proof of Bougerol's Identity in Law

As we remarked in the introduction, Bougerol's identity (1.d) is equivalent to (1.e). In order to prove formula (1.e), it is sufficient to show that both sides admit the same Laplace transform in t; hence, all we need to prove is the following: for any $\theta \geq 0$,

$$\int_0^\infty \frac{dt}{\sqrt{(1+u^2)t}} \exp\left(-\frac{\theta^2 t}{2}\right) \exp\left(-\frac{1}{2t}(\text{Arg sh } u)^2\right)$$

$$= \int_0^\infty dt \, \exp\left(-\frac{\theta^2 t}{2}\right) E\left[\frac{1}{\sqrt{A_t}} \exp\left(-\frac{u^2}{2A_t}\right)\right] \tag{3.a}$$

holds. We denote by $l(\theta, u)$, the left-hand side and by $r(\theta, u)$, the right-hand side of (3.a).

We first transform $r(\theta, u)$ using the expression of $(\exp(B_t), t \geq 0)$ given in (2.a) in terms of $(\rho_u, u \geq 0)$, a two-dimensional Bessel process starting from 1.

We then obtain, with the notation

$$H_s = \int_0^s \frac{du}{\rho_u^2},$$

$$r(\theta, u) = E\left[\int_0^\infty dH_s \exp\left(-\frac{\theta^2 H_s}{2}\right) \frac{1}{\sqrt{s}} \exp\left(-\frac{u^2}{2s}\right)\right]$$

$$= E\left[\int_0^\infty \frac{ds}{\sqrt{s}} \frac{1}{\rho_s^2} \exp\left(-\frac{\theta^2 H_s}{2}\right) \exp\left(-\frac{u^2}{2s}\right)\right]. \tag{3.b}$$

We now replace $\exp(-\theta^2 H_s/2)$ by its conditional expectation, given ρ_s, so that, using the formulae (2.e), and then (2.d), we obtain

$$r(\theta, u) = \int_0^\infty \frac{ds}{\sqrt{s}} \left(\frac{1}{s}\right) \int_0^\infty \frac{d\rho \rho}{\rho^2} \exp\left(-\frac{1+\rho^2}{2s}\right) I_\theta\left(\frac{\rho}{s}\right) \exp\left(-\frac{u^2}{2s}\right),$$

$$= \int_0^\infty \frac{d\rho}{s^{\frac{3}{2}}} \int_0^\infty \frac{d\rho}{\rho} \exp\left(-\frac{1+\rho^2+u^2}{2s}\right) I_\theta\left(\frac{\rho}{s}\right),$$

$$= \int_0^\infty \frac{d\rho}{\rho} \int_0^\infty \frac{ds}{s^{\frac{3}{2}}} \exp\left(-\frac{1+\rho^2+u^2}{2s}\right) I_\theta\left(\frac{\rho}{s}\right). \tag{3.c}$$

We now define

$$M(\rho, u) = \int_0^\infty \frac{ds}{s^{\frac{3}{2}}} \exp\left(-\frac{1+\rho^2+u^2}{2s}\right) I_\theta\left(\frac{\rho}{s}\right).$$

Making the change of variables $s = \rho/\xi$, we obtain

$$M(\rho, u) = \frac{1}{\sqrt{\rho}} \int_0^\infty \frac{d\xi}{\sqrt{\xi}} \exp\left(-\left(\frac{1+\rho^2+u^2}{2\rho}\right)\xi\right) I_\theta(\xi),$$

so that, using Fubini's theorem in (3.c), we obtain

$$r(\theta, u) = \int_0^\infty \frac{d\xi}{\sqrt{\xi}} N(u, \xi) I_\theta(\xi), \tag{3.d}$$

where

$$N(u, \xi) = \int_0^\infty \frac{d\rho}{\rho^{\frac{3}{2}}} \exp\left(-\frac{1}{2}\left(\frac{1+u^2}{\rho} + \rho\right)\xi\right).$$

Making some elementary changes of variables, we obtain, using (2.g) and (2.g′) for $\nu = \frac{1}{2}$:

$$N(u, \xi) = \frac{\sqrt{2\pi}}{\xi} \frac{1}{\sqrt{1+u^2}} \exp(-\xi\sqrt{1+u^2}).$$

Consequently, we deduce from (3.d), with the help of (2.f), that

$$r(\theta, u) = \frac{\sqrt{2\pi}}{\sqrt{1+u^2}} \frac{1}{\theta} \frac{1}{(\sqrt{1+u^2}+u)^\theta}.$$

On the other hand, the left-hand side $l(\theta, u)$ of (3.a) is easily seen, thanks to the formulae (2.g) and (2.g'), for $\nu = -\frac{1}{2}$, to be equal to

$$l(\theta, u) = \frac{\sqrt{2\pi}}{\sqrt{1+u^2}} \frac{1}{\theta} \exp{-\theta(\text{Arg sh } u)},$$

which is equal to $r(\theta, u)$, since Arg sh $u \equiv \log(u + \sqrt{1+u^2})$.

4. The Moments of A_t

4.1. Elementary arguments, which do not bear upon the previous section, and which involve essentially the independence of the increments of B, allow us to obtain an explicit formula for

$$E\left[\left(\int_0^t ds \exp(B_s)\right)^n \exp(\mu B_t)\right],$$

for any $n \in \mathbb{N}$, and $\mu \geq 0$.

In order to simplify the presentation, and to be able to extend some of the computations for the Brownian case to the case of some other processes with independent increments, we shall write for $\lambda \in \mathbb{R}$,

$$E[\exp(\lambda B_t)] = \exp t\varphi(\lambda), \quad \text{where, here, } \varphi(\lambda) = \frac{\lambda^2}{2}. \tag{4.a}$$

We then have the following result.

Theorem 1. 1. *Let $\mu \geq 0$, $n \in \mathbb{N}$, and $\alpha > \varphi(\mu + n) \equiv \frac{1}{2}(\mu + n)^2$. Then the formula*

$$\int_0^\infty dt \exp(-\alpha t) E\left[\left(\int_0^t ds \exp B_s\right)^n \exp(\mu B_t)\right] = \frac{n!}{\displaystyle\prod_{j=0}^n (\alpha - \varphi(\mu + j))}$$

$$\tag{4.b}$$

holds.
2. *Let $\mu \geq 0$, $n \in \mathbb{N}$, and $t \geq 0$. Then we have*[1]

$$E\left[\left(\int_0^t ds \exp(B_s)\right)^n \exp(\mu B_t)\right] = E[P_n^{(\mu)}(\exp B_t) \exp(\mu B_t)] \tag{4.c}$$

---

[1] See also Postscript #3.

where $(P_n^{(\mu)}, n \in \mathbb{N})$ is the following sequence of polynomials:

$$P_n^{(\mu)}(z) = n! \left(\sum_{j=0}^{n} c_j^{(\mu)} z^j \right), with \ c_j^{(\mu)} = \prod_{\substack{k \neq j \\ 0 \leq k \leq n}} (\varphi(\mu + j) - \varphi(\mu + k))^{-1}.$$

Proof.   1. We define

$$\Phi_{n,t}(\mu) = E \left[\left(\int_0^t ds \exp(B_s) \right)^n \exp(\mu B_t) \right]$$

$$= n! E \left[\int_0^t ds_1 \int_0^{s_1} ds_2 \ldots \int_0^{s_{n-1}} ds_n \exp(B_{s_1} + \cdots + B_{s_n} + \mu B_t) \right].$$

We then remark that

$$E[\exp(\mu B_t + B_{s_1} + \cdots + B_{s_n})]$$
$$= E[\exp\{\mu(B_t - B_{s_1}) + (\mu + 1)(B_{s_1} - B_{s_2}) + \cdots + (\mu + n)B_{s_n}\}]$$
$$= \exp\{\varphi(\mu)(t - s_1) + \varphi(\mu + 1)(s_1 - s_2) + \cdots + \varphi(\mu + n)s_n\}.$$

Therefore, we have

$$\int_0^\infty dt \exp(-\alpha t) \Phi_{n,t}(\mu)$$

$$= n! \int_0^\infty dt \exp(-\alpha t) \int_0^t ds_1 \int_0^{s_1} ds_2$$

$$\ldots \int_0^{s_{n-1}} ds_n \exp\{\varphi(\mu)(t - s_1) + \cdots + \varphi(\mu + n)s_n\}$$

$$= n! \int_0^\infty ds_n \exp -(\alpha - \varphi(\mu + n))s_n$$

$$\ldots \int_{s_n}^\infty ds_{n-1} \exp -(\alpha - \varphi(\mu + n - 1))(s_{n-1} - s_n)$$

$$\ldots \int_{s_1}^\infty dt \exp -(\alpha - \varphi(\mu))(t - s_1),$$

so that, in the case $\alpha > \varphi(\mu + n)$, we obtain formula (4.b) by integrating successively the $(n + 1)$ exponential functions.

2. Next, we use the additive decomposition formula

$$\frac{1}{\prod\limits_{j=0}^{n} (\alpha - \varphi(\mu + j))} = \sum_{j=0}^{n} c_j^{(\mu)} \frac{1}{(\alpha - \varphi(\mu + j))}$$

where $c_j^{(\mu)}$ is given as stated in the theorem, and we obtain, for $\alpha > \varphi(\mu + n)$:

$$\int_0^\infty dt \exp(-\alpha t)\Phi_{n,t}(\mu) = n! \sum_{j=0}^n c_j^{(\mu)} \int_0^\infty dt \exp(-\alpha t) \exp(\varphi(\mu + j)t),$$

a formula from which we deduce

$$\Phi_{n,t}(\mu) = n! \sum_{j=0}^n c_j^{(\mu)} \exp(\varphi(\mu + j)t) = n! \sum_{j=0}^n c_j^{(\mu)} E[\exp(jB_t) \exp(\mu B_t)]$$

$$= E[P_n^{(\mu)}(\exp B_t) \exp(\mu B_t)].$$

Hence, we have proved formula (4.c). $\qquad\square$

As a consequence of Theorem 1, we have the following.

Corollary 1. *For any $\lambda \in \mathbb{R}$, and any $n \in \mathbb{N}$, we have*

$$\lambda^{2n} E\left[\left(\int_0^t du \exp(\lambda B_u)\right)^n\right] = E[P_n(\exp \lambda B_t)] \qquad (4.d)$$

where

$$P_n(z) = 2^n(-1)^n \left\{\frac{1}{n!} + 2\sum_{j=1}^n \frac{n!(-z)^j}{(n-j)!(n+j)!}\right\}. \qquad (4.e)$$

Proof. Thanks to the scaling property of Brownian motion, it suffices to prove formula (4.d) for $\lambda = 1$, and any $t \geq 0$. In this case, we remark that formula (4.d) is then precisely formula (4.c) taken with $\mu = 0$, once the coefficients $c_j^{(0)}$ have been identified as

$$c_0^{(0)} = (-1)^n \frac{2^n}{(n!)^2}; \quad c_j^{(0)} = \frac{2^n(-1)^{n-j}2}{(n-j)!(n+j)!} \ (1 \leqq j \leqq n);$$

therefore, it now appears that the polynomial P_n is precisely $P_n^{(0)}$, and this ends the proof of (4.d). $\qquad\square$

It may also be helpful to write down explicitly the moments of $A_t^{(\nu)}$.

Corollary 2. *For any $\lambda \in \mathbb{R}^*, \mu \in \mathbb{R}$, and $n \in \mathbb{N}$, we have*

$$\lambda^{2n} E\left[\left(\int_0^t du \exp \lambda(B_u + \mu u)\right)^n\right] = n! \sum_{j=0}^n c_j^{(\mu/\lambda)} \exp\left(\left(\frac{\lambda^2 j^2}{2} + \lambda j\mu\right)t\right).$$

$$(4.d')$$

In particular, we have, for $\mu = 0$,

$$\lambda^{2n} E\left[\left(\int_0^t du \exp \lambda B_u\right)^n\right]$$

$$= n!\left\{\frac{(-1)^n}{(n!)^2} + 2\sum_{j=1}^n \frac{(-1)^{n-j}}{(n-j)!(n+j)!} \exp \frac{\lambda^2 j^2 t}{2}\right\}.$$

(4.d'')

Taking $\lambda = 2$ in formula (4.d''), we obtain

$$E[A_t^n] = \frac{1}{4^n} E[P_n(\exp 2B_t)].$$

(4.f)

A different computation of $E[A_t^n]$ follows from Bougerol's identity:

$$\text{for any } t \geq 0, \qquad \gamma_{A_t} \overset{(law)}{=} \sinh(B_t),$$

(1.d)

from which we deduce

$$E[\gamma_1^{2n}]E[A_t^n] = E[(\sinh(B_t))^{2n}].$$

(4.g)

We now show that formulae (4.f) and (4.g) give the same value for $E[A_t^n]$, thereby providing a partial checking of Theorem 1 and its corollary.

Lemma 1. *The identity*

$$\frac{1}{4^n} E[P_n(\exp 2B_t)] = \frac{E[(\sinh B_t)^{2n}]}{E(B_1^{2n})}$$

(4.h)

holds for all $n \in \mathbb{N}$ and $t \geq 0$.

Proof. We write g_{2n} for $E(B_1^{2n})$. Developing both sides of (4.h) with the help of the distribution of B_t, we see that (4.h) is equivalent to

$$\frac{1}{4^n}\int_{-\infty}^\infty dx \exp\left(-\frac{x^2}{2t}\right) P_n(\exp 2x) = \frac{1}{g_{2n}}\int_{-\infty}^\infty dx \exp\left(-\frac{x^2}{2t}\right)(\sinh x)^{2n},$$

which is itself equivalent to

$$\int_0^\infty dx \exp(-x^2/2t)(P_n(\exp(2x)) + P_n(\exp(-2x)))$$

$$= \frac{4^n}{g_{2n}} 2 \int_0^\infty dx \exp(-x^2/2t)(\sinh x)^{2n}.$$

(4.h')

From the injectivity of the Laplace transform, we deduce that (4.h') is satisfied for all $t > 0$ if, and only if

$$\frac{1}{2}(P_n(\exp(2x)) + P_n(\exp(-2x))) = \frac{4^n}{g_{2n}}(\sinh x)^{2n},$$

or, equivalently, if we use the variable $y = \exp(2x)$,

$$\frac{1}{2}\left(P_n(y) + P_n\left(\frac{1}{y}\right)\right) = \frac{1}{g_{2n}}\left(y + \frac{1}{y} - 2\right)^{2n}. \tag{4.i}$$

We now remark that

$$\left(y + \frac{1}{y} - 2\right)^n \equiv \left(\sqrt{y} - \frac{1}{\sqrt{y}}\right)^{2n} \stackrel{\text{def}}{=} Q_n(y)$$

is given by the formula

$$Q_n(y) = \sum_{k=0}^{2n} C_{2n}^k y^{n-k}(-1)^k. \tag{4.j}$$

We now write

$$Q_n(y) = \sum_{k=0}^{n-1} C_{2n}^k y^{n-k}(-1)^k + (-1)^n C_{2n}^n + \sum_{k=n+1}^{2n} C_{2n}^{n-(k-n)}\frac{(-1)^{n-(k-n)}}{y^{k-n}}$$

$$= (-1)^n\left\{C_{2n}^n + \sum_{k=1}^{n} C_{2n}^{n-k}(-y)^k + \sum_{k=1}^{n} C_{2n}^{n-k}\left(-\frac{1}{y}\right)^k\right\}. \tag{4.j$'$}$$

On the other hand, the formula $g_{2n} = (2n)!/2^n n!$, which follows from $E[\exp(\lambda B_1)] = \exp(\frac{1}{2}\lambda^2)$ by differentiation, is well known. Hence, the formulae (4.j$'$) and (4.e) imply that

$$\frac{1}{g_n}Q_n(y) = \frac{1}{2}\left(P_n(y) + P_n\left(\frac{1}{y}\right)\right),$$

which is precisely (4.i). □

4.2. Although we do not want to sidetrack the reader with questions of secondary importance, it may be worthwhile to mention that the results of Theorem 1 can be extended to a large class of processes $(X_t, t \geq 0)$ with independent increments, instead of $(B_t, t \geq 0)$. We hope to develop such a general study in a further paper, but we may remark here that, for $X_t = B_t^{(\nu)} \equiv B_t + \nu t$, with $\nu \geq 0$, formulae (4.b) and (4.c) may be extended with the function $\varphi^{(\nu)}$ now defined to be $\varphi^{(\nu)}(\lambda) = \frac{1}{2}\lambda^2 + \nu\lambda$. In particular, for $\mu, \nu \geq 0$, there exist polynomials $^{(\nu)}P_n^{(\mu)}$ such that

$$E\left[\left(\int_0^t ds\,\exp(B_s^{(\nu)})\right)^n \exp(\mu B_t^{(\nu)})\right]$$

$$= E[^{(\mu)}P_n^{(\nu)}(\exp B_t^{(\nu)})\exp(\mu B_t^{(\nu)})]. \tag{4.c$_\nu$}$$

Now, thanks to Girsanov's theorem, the above identity is equivalent to

$$E\left[\left(\int_0^t ds \exp(B_s)\right)^n \exp((\mu + \nu)B_t)\right]$$

$$= E[^{(\nu)}P_n^{(\mu)}(\exp B_t)\exp((\mu + \nu)B_t)]$$

(4.c)$'_\nu$

and, comparing this with formula (4.c), we easily obtain the following.

Lemma 2. *For any $n \in \mathbb{N}$, $\mu, \nu \geq 0$, we have $^{(\nu)}P_n^{(\mu)} = P_n^{(\mu+\nu)}$.*

Proof. Comparing formulae (4.c)$'_\nu$ and (4.c) with the parameter μ replaced there by $(\mu + \nu)$, we obtain

$$E[^{(\nu)}P_n^{(\mu)}(\exp B_t)] = E[P_n^{(\mu+\nu)}(\exp B_t)\exp(\mu + \nu)B_t].$$

Denote $\lambda = \mu + \nu$. Now, a Laplace transform argument (in t) shows that the preceding equality is true if and only if

$$P(\rho)\rho^\lambda + P\left(\frac{1}{\rho}\right)\left(\frac{1}{\rho}\right)^\lambda = Q(\rho)\rho^\lambda + Q\left(\frac{1}{\rho}\right)\left(\frac{1}{\rho}\right)^\lambda, \qquad \rho > 0$$

where we have written P for $^{(\nu)}P_n^{(\mu)}$ and Q for $P_n^{(\mu+\nu)}$.

Now, letting $\rho \to \infty$, we obtain that the coefficients of highest degree (i.e. n) of P and Q are equal. Iterating this procedure, we show that the coefficients of any given degree $k(\leq n)$ of P and Q are equal, hence $P = Q$. $\square$

Remark. The identity shown in Lemma 2 could also be proved directly upon looking at the explicit formulae for $^{(\nu)}P_n^{(\mu)}$ and $P_n^{(\mu+\nu)}$ (see just below (4.c)). Indeed, it suffices to remark that

$$\varphi^{(\nu)}(\mu + x) - \varphi^{(\nu)}(\mu + y) \equiv (\mu + \nu)(x - y) = \varphi^{(\mu+\nu)}(x) - \varphi^{(\mu+\nu)}(y).$$

However, it also seemed interesting to develop the arguments of the above proof, which are not computational.

4.3. We now remark that the polynomials $P_n \equiv P_n^{(0)}$ are closely linked with certain hypergeometric polynomials. We first recall the definition of the hypergeometric function F with parameters α, β, γ:

$$F(\alpha, \beta, \gamma; z) = \sum_{z=0}^{\infty} \frac{(\alpha)_k(\beta)_k}{(\gamma)_k \, k!} z^k,$$

where

$$(\lambda)_k = \frac{\Gamma(\lambda + k)}{\Gamma(\lambda)} \equiv \lambda(\lambda + 1) \cdots (\lambda + k - 1)$$

(see Lebedev [8], p. 238). In particular, we have

$$F(-n, 1, n + 1; z) = \sum_{j=0}^{n} \frac{(-n)_j (1)_j}{(n+1)_j} \frac{z^j}{j!} = \sum_{j=0}^{n} \frac{(n!)^2 (-z)^j}{(n+j)!(n-j)!}.$$

As a consequence, we deduce from formula (4.e) the following relation:

$$P_n(z) = \frac{(-2)^n}{n!} \{2F(-n, 1, n + 1; z) - 1\}. \tag{4.k}$$

4.4. Although in Section 4.1 above, we proved the identity

$$E[\gamma_1^{2n}] E[A_t^n] = E[(\sinh B_t)^{2n}], \text{ for any } n \in \mathbb{N}, \tag{4.g}$$

it does not seem possible to deduce Bougerol's identity in law (1.d) directly from (4.g). Indeed, the law of $\sinh(B_t)$ is not determined by its moments, and, in fact, it is easy to exhibit some explicit distributions on $\mathbb{R}$ which have the same moments as $\sinh(B_t)$.

For simplicity, we take $t = 1$, and we recall (see Stoyanov [11], p. 89) that if N is a standard $\mathcal{N}(0, 1)$ random variable, then, for any ε with $0 < |\varepsilon| \leq 1$, and any $p \in \mathbb{Z}^*$, $\exp(N)$ and $\exp(N_{(\varepsilon,p)})$ have the same positive and negative moments, i.e.

$$E[\exp(kN)] = E[\exp(kN_{(\varepsilon,p)})], \quad \text{for any } k \in \mathbb{Z},$$

if $P(N_{(\varepsilon,p)} \in dx) \overset{\text{def}}{=} P(N \in dx)(1 + \varepsilon \sin(\pi px))$. Consequently, $\sinh(N)$ and $\sinh(N_{(\varepsilon,p)})$ also have the same positive and negative moments, although they have different distributions. However, we have not been able to show that $\sinh(N_{(\varepsilon,p)})$ can be represented in law as $\gamma_{A_{(\varepsilon,p)}}$, for some non-negative random variable $A_{(\varepsilon,p)}$ which is independent of the Brownian motion $(\gamma_t, t \geq 0)$. Hence, the possibility of deducing (1.d) directly from (4.g) is not entirely ruled out, although it seems very unlikely.

4.5. Coming back to (4.c) and (4.c)$_\nu$, we give yet another example of such formulae, this time for the standard Cauchy process $(C_t, t \geq 0)$, the law of which, as a process with homogeneous independent increments, is determined by

$$E[\exp(i\lambda C_t)] = \exp(-t |\lambda|) \quad (t \geq 0, \lambda \in \mathbb{R}).$$

Computations similar to those developed in the proof of Theorem 1 yield the following very simple formula: for any $t \geq 0$, and any $\mu \geq 0$,

$$E\left[\left(\int_0^t ds \, \exp\,(iC_s)\right)^n \exp\,(i\mu C_t)\right] = E[(1 - \exp\,(iC_t))^n \exp\,(i\mu C_t)].$$

(4.1)

5. The Law of $A^{(\nu)}$ Taken at an Independent Exponential Time

5.1. We now consider S_θ, an exponentially distributed random variable with parameter $\frac{1}{2}\theta^2$, that is, $P(S_\theta \in dt) = \frac{1}{2}\theta^2 \exp\,(-\frac{1}{2}\theta^2 t) \, dt$, which is assumed moreover to be independent of B.

A simple variant of the arguments used in Section 3 above will yield the following result.

Theorem 2. *We recall that, for $\mu \geq 0$, $p_t^\mu(a, \rho) \, d\rho$ denotes the semigroup of the Bessel process of index μ, which is given by formula (2.d).*

1. *The joint law of $(\exp\,(B_{S_\theta}), A_{S_\theta})$ is given by*

$$P(\exp\,(B_{S_\theta}) \in d\rho; A_{S_\theta} \in du) = \frac{\theta^2}{2\rho^{2+\theta}} \, p_u^\theta(1, \rho) \, d\rho \, du.$$

(5.a)

2. *More generally, if $\nu \in \mathbb{R}$, we set: $\lambda = (\theta^2 + \nu^2)^{1/2}$, and we have*

$$P(\exp\,(B_{S_\theta}^{(\nu)}) \in d\rho; A_{S_\theta}^{(\nu)} \in du) = \frac{\theta^2}{2\rho^{2+\lambda-\nu}} \, p_u^\lambda(1, \rho) \, d\rho \, du.$$

(5.b)

Proof.   1. Consider $f, g : \mathbb{R}_+ \to \mathbb{R}_+$ two Borel functions. We have

$$\Phi_\theta(f, g) \stackrel{\text{def}}{=} E[f(\exp B_{S_\theta})g(A_{S_\theta})]$$

$$= \frac{\theta^2}{2} E\left[\int_0^\infty dt \, \exp\left(-\frac{\theta^2 t}{2}\right) f(\exp B_t)g(A_t)\right].$$

Now, using the formulae (2.a), (2.e) and (2.d), we obtain

$$\Phi_\theta(f, g) = \frac{\theta^2}{2} E\left[\int_0^\infty dH_u \, \exp\left(-\frac{\theta^2 H_u}{2}\right) f(\exp B_{H_u})g(u)\right]$$

$$= \frac{\theta^2}{2}\left[\int_0^\infty \frac{du}{\rho_u^2} \exp\left(-\frac{\theta^2 H_u}{2}\right) f(\rho_u)g(u)\right]$$

$$= \frac{\theta^2}{2}\int_0^\infty du \, g(u) \int_0^\infty \frac{d\rho}{\rho^{2+\theta}} f(\rho)p_u^\theta(1, \rho),$$

which implies formula (5.a).

2. More generally, we have

$$\Phi_\theta^{(\nu)}(f,g) \overset{\text{def}}{=} E[f(\exp(B_{S_\theta}^{(\nu)}))g(A_{S_\theta}^{(\nu)})]$$

$$= \frac{\theta^2}{2}\int_0^\infty dt \exp\left(-\frac{\theta^2 t}{2}\right) E[f(\exp B_t^{(\nu)})g(A_t^{(\nu)})]$$

$$= \frac{\theta^2}{2}\int_0^\infty dt \exp\left(-\frac{\theta^2 t}{2}\right) E\left[f(\exp B_t)g(A_t)\exp\left(\nu B_t - \frac{\nu^2 t}{2}\right)\right]$$

$$= \frac{\theta^2}{(\theta^2+\nu^2)}\left(\frac{\theta^2+\nu^2}{2}\right)\int_0^\infty dt \exp\left(-\frac{\lambda^2 t}{2}\right)\ldots$$

$$\ldots E[f(\exp B_t)g(A_t)\exp(\nu B_t)] \qquad (5.c)$$

$$= \frac{\theta^2}{(\theta^2+\nu^2)}E[f(\exp B_{S_\lambda})g(A_{S_\lambda})\exp(\nu B_{S_\lambda})]$$

$$= \frac{\theta^2}{2}\int_0^\infty du\, g(u)\int_0^\infty \frac{d\rho}{\rho^{\lambda+2}}\rho^\nu f(\rho)p_u^\lambda(1,\rho), \quad \text{from (5.a)},$$

which implies formula (5.b). Note that in (5.c), we have used Girsanov's theorem to relate the law of $(B_u^{(\nu)}, u \le t)$ to that of $(B_u, u \le t)$. □

From formula (5.b), we immediately obtain the conditional law of $A_{S_\theta}^{(\nu)}$, given $B_{S_\theta}^{(\nu)}$.

Corollary. *We keep the notation used in Theorem 2. We have, for any $\nu \in \mathbb{R}$,*

$$P(B_{S_\theta}^{(\nu)} \in dx) = \frac{\theta^2}{2\lambda}\exp(-\lambda|x|+\nu x)\, dx, \qquad (5.d)$$

and

$$P(A_{S_\theta}^{(\nu)} \in du \mid B_{S_\theta}^{(\nu)} = x) = \lambda\left(\exp(-x+2\lambda x^-)\right)p_u^\lambda(1,e^x)du \qquad (5.d')$$

where $x^- = (-x)1_{(x \le 0)}$.

Proof.   The formula (5.d) may be obtained using Girsanov's theorem to reduce it to the case $\nu = 0$, in which case we find the well-known result

$$P(B_{S_\theta} \in dx) = \frac{\theta}{2}\exp(-\theta|x|)\, dx.$$

Once formula (5.d) has been obtained, formula (5.d') follows from (5.b). □

5.2. We now discuss the statements of Theorem 2 and its corollary in relation to the explicit description of the laws of last-passage times for certain transient diffusions on $\mathbb{R}_+$, and, in particular, Bessel processes with dimension $d > 2$.

The following general discussion may be found in Pitman and Yor [9], Section 6, Williams [14], Revuz and Yor [10] (Exercises (4.16), p. 298, and (1.16), p. 378).

We consider the canonical realization on $C(\mathbb{R}_+, \mathbb{R}_+)$ of a regular diffusion $(R_t, t \geq 0; P_x, x \in \mathbb{R}_+)$ with infinite lifetime, and we suppose for simplicity that

$$P_x(T_0 < \infty) = 0, \quad x > 0, \tag{5.e}$$

$$P_x\left(\lim_{t \to \infty} R_t = \infty\right) = 1, \quad x > 0. \tag{5.f}$$

As a consequence of (5.e) and (5.f), a scale function s for this diffusion satisfies $s(0+) = -\infty$ and $s(\infty) < \infty$. We can therefore suppose that $s(\infty) = 0$. Let Γ be the infinitesimal generator of the diffusion, and take the speed measure m to be such that $\Gamma = \frac{1}{2}(d/dm)(d/ds)$. According to Itô and McKean ([5], p. 149), there exists a continuous function

$$p^* : \begin{array}{c} (\mathbb{R}_+^*)^3 \to \mathbb{R}_+^* \\ (t, x, y) \to p_t^*(x, y) \end{array}$$

which is symmetric in x and y, and such that the semigroup P_t of the diffusion is given by

$$P_t(x, dy) = p_t^*(x, y)m(dy).$$

We can now state the following result.

Theorem 3. (*Pitman and Yor [9], Theorem 6.1*). *Let (R_t, P_x) be a regular diffusion on $\mathbb{R}_+$, which satisfies the hypotheses (5.e) and (5.f). Then, denoting $T_a = \inf\{t : R_t = a\}$, and $L_b = \sup\{t : R_t = b\}$,*

1. *for all $a, b > 0$,*

$$P_a(L_b \in dt) = \frac{-1}{2s(b)} p_t^*(a, b) dt \quad (t > 0), \tag{5.g}$$

2. *for $a \leq b$, the formula (5.g) defines an infinitely divisible probability distribution on $\mathbb{R}_+$,*

3. *for $a < b$, we have*

$$P_b(T_a < \infty) \equiv P_b(L_a > 0) = \frac{s(b)}{s(a)}$$

and $\quad P_b(L_a \in dt; L_a > 0) = \frac{-1}{2s(a)} p_t^*(b, a)\, dt,$

so that

$$P_b(L_a \in dt \mid L_a > 0) = \frac{-1}{2s(b)} p_t^*(a, b)\, dt = P_a(L_b \in dt).$$

As a consequence of Theorem 3, we may now interpret, after making some elementary computations concerning the speed measures and scale functions of Bessel processes, the corollary of Theorem 2 as follows.

Proposition 1. *We keep the notation used in Theorem 2. Moreover, we denote by $L^{(\lambda)}(a, b)$ the last-passage time in b of the Bessel process $\rho^{(\lambda)}$ starting from a. Then, for any $\nu \in \mathbb{R}$, the quantity*

$$P(A^{(\nu)}_{S_\theta} \in du \mid B^{(\nu)}_{S_\theta} = x)$$

is equal to

$$P(L^{(\lambda)}(1, e^x) \in du) \quad if \quad x \geqq 0,$$

and to

$$P(L^{(\lambda)}(1, e^x) \in du \mid L^{(\lambda)}(1, e^x) > 0) = P(L^{(\lambda)}(e^x, 1) \in du) \quad if \quad x < 0.$$

5.3. We now show how the results in Theorem 2 and its corollary, together with Proposition 1, lead to the following result:

$$\text{for } \nu > 0, \quad A^{(-\nu)}_\infty \overset{\text{(law)}}{\underset{(a)}{=}} \frac{1}{2Z_\nu} \overset{\text{(law)}}{\underset{(b)}{=}} L^{(\nu)}(0, 1), \tag{5.h}$$

where Z_ν denotes a standard gamma variable with parameter ν, i.e.

$$P(Z_\nu \in dt) = \frac{t^{\nu-1}}{\Gamma(\nu)} \exp(-t) \, dt \quad (t > 0)$$

(for a more direct derivation of (5.h) and related references, see [16]). Since $S_\theta \overset{(P)}{\underset{\theta \to 0}{\longrightarrow}} \infty$, it will be sufficient, in order to prove (5.h), to pass to the limit as $\theta \to 0$, in formulae (5.d) and (5.d'). In fact, writing $\mu = 1/2\nu$, we shall show

$$(-\theta^2 B^{(-\nu)}_{S_\theta}; A^{(-\nu)}_{S_\theta}) \overset{\text{(law)}}{\underset{\theta \to 0}{\longrightarrow}} \left(\mathbf{e}_\mu; \frac{1}{2Z_\nu} \right) \tag{5.h'}$$

where $\mathbf{e}_\mu$ and Z_ν are independent, and $P(\mathbf{e}_\mu \in dy) = 1/2\nu \, \exp(-(y/2\nu)) \, dy$ $(y > 0)$. The convergence

$$-\theta^2 B^{(-\nu)}_{S_\theta} \overset{\text{(law)}}{\underset{\theta \to 0}{\longrightarrow}} \mathbf{e}_\mu$$

is immediate, since

$$-\theta^2 B^{(-\nu)}_{S_\theta} \overset{\text{(law)}}{=} (-\theta^2)(\sqrt{S_\theta} B_1 - \nu S_\theta), \text{ and} \quad \frac{\theta^2}{2} S_\theta \overset{\text{(law)}}{=} \mathbf{e}_1.$$

Then, in order to prove ((5.h), (a)), it suffices to show that, for any function $f : \mathbb{R}_+ \to \mathbb{R}_+$, continuous, with compact support, and any $y > 0$, the quantity

$$E\left[f(A_{S_\theta}^{(-\nu)}) \mid B_{S_\theta}^{(-\nu)} = -\frac{y}{\theta^2}\right]$$

$$\equiv \lambda \exp\left\{\frac{y}{\theta^2}(1 + 2\lambda)\right\} \int_0^\infty du \, p_u^\lambda\left(1, \exp\left(-\frac{y}{\theta^2}\right)\right) f(u) \qquad \text{(5.d'')}$$

converges, as $\theta \to 0$, towards

$$\int_0^\infty \frac{du}{\Gamma(\nu)2^\nu u^{\nu+1}} \exp\left(-\frac{1}{2u}\right) f(u).$$

This is a consequence of the explicit formula (2.d) for $p_u^\lambda(1, \xi)$ and of the well-known equivalence result

$$I_\lambda(\xi) \sim \left(\frac{\xi}{2}\right)^\lambda \frac{1}{\Gamma(\lambda+1)} \qquad \text{as } \xi \to 0,$$

uniformly as λ varies in a compact subset of $\mathbb{R}_+^*$. On the other hand, we shall prove ((5.h), (b)), or rather the identity

$$A_\infty^{(-\nu)} \overset{\text{(law)}}{=} L^{(\nu)}(0, 1) \qquad \text{(5.i)}$$

with the help of Proposition 1. Indeed, from this proposition, we have, for any $y > 0$, and any bounded continuous function $f : \mathbb{R}_+ \to \mathbb{R}_+$,

$$E\left[f(A_{S_\theta}^{(-\nu)}) \mid B_{S_\theta}^{(-\nu)} = -\frac{y}{\theta^2}\right]$$

$$= E\left[f\left(L^{(\lambda)}\left(1, \exp\left(-\frac{y}{\theta^2}\right)\right)\right) \mid L^{(\lambda)}\left(1, \exp\left(-\frac{y}{\theta^2}\right)\right) > 0\right]$$

$$= E\left[f\left(L^{(\lambda)}\left(\exp\left(-\frac{y}{\theta^2}\right), 1\right)\right)\right]$$

and, finally, letting $\theta \to 0$, the last-written quantity converges to $E[f(L^{(\nu)}(0, 1))]$ thereby proving (5.i).

6.   The Law of A Taken at a Fixed Time

6.1. Our aim in this section is to give a formula, which we would like to be as explicit as possible, for

$$P(A(t) \in du \mid B_t = x) \overset{\text{def}}{=} a_t(x, u) \, du. \qquad \text{(6.a)}$$

This computation will be closely linked with the following form of the density of the Hartman–Watson distribution $\eta_r(du)$, which we defined above in (2.e'). We take the following formula from Yor ([15], p. 85):

$$I_0(r)\eta_r(du) = \theta_r(u) \, du, \quad \text{with} \quad \theta_r(u) = \frac{r}{(2\pi^3 u)^{\frac{1}{2}}} \exp\left(\frac{\pi^2}{2u}\right) \psi_r(u), \qquad \text{(6.b')}$$

and

$$\psi_r(u) = \int_0^\infty dy \exp\left(-y^2/2u\right) \exp\left(-r(\cosh y)\right)(\sinh y) \sin\left(\frac{\pi y}{u}\right). \quad (6.b'')$$

Before getting any further into the computation of $a_t(x, u)$, we remark that, thanks to Girsanov's theorem relating the laws of $(B_s + \nu s; s \leq t)$ and $(B_s; s \leq t)$, we have

$$P(A_t^{(\nu)} \in du \mid B_t + \nu t = x) = a_t(x, u)\, du \quad (6.a')$$

so that it is really sufficient to consider only the case $\nu = 0$.

6.2. The main result in this section is the following.

Proposition 2. *If we denote $P(A_t \in du | B_t = x) = a_t(x, u)\, du$, then we have*

$$\frac{1}{\sqrt{2\pi t}} \exp\left(-\frac{x^2}{2t}\right) a_t(x, u) = \frac{1}{u} \exp\left(-\frac{1}{2u}(1 + e^{2x})\right) \theta_{e^x/u}(t). \quad (6.c)$$

Proof.   Consider $f, g : \mathbb{R}_+ \to \mathbb{R}_+$ two Borel functions; then, we have, on the one hand, for $\mu \geq 0$:

$$E\left[\int_0^\infty dt \exp\left(-\frac{\mu^2 t}{2}\right) f(\exp B_t) g(A_t)\right]$$
$$= \int_0^\infty dt \exp\left(-\frac{\mu^2 t}{2}\right) \int_{-\infty}^\infty \frac{dx}{\sqrt{2\pi t}} f(e^x) \exp\left(-\frac{x^2}{2t}\right) \int_0^\infty du\, g(u) a_t(x, u),$$

by definition of $a_t(x, u)$; on the other hand, the same quantity is, by formula (5.a), equal to

$$\int_0^\infty du\, g(u) \int_0^\infty \frac{d\rho}{\rho^{\mu+2}} f(\rho) p_u^\mu(1, \rho)$$
$$= \int_{-\infty}^\infty dx \exp(-(\mu+1)x) f(e^x) \int_0^\infty du\, g(u) p_u^\mu(1, e^x).$$

Comparing the two expressions we have just obtained, we see that

$$\int_0^\infty dt \exp-\frac{1}{2}\left(\mu^2 t + \frac{x^2}{t}\right) a_t(x, u) = \exp(-(\mu+1)x) p_u^\mu(1, e^x). \quad (6.d)$$

Now, using the explicit form of p^μ given by formula (2.d), and the definition (6.b) of the function θ given above, the right-hand side of (6.d) may be written as

$$\frac{1}{u} \exp\left(-\frac{1}{2u}(1 + e^{2x})\right) \int_0^\infty dt\, \theta_{e^x/u}(t) \exp\left(-\frac{\mu^2 t}{2}\right),$$

and, finally, formula (6.c) follows from (6.d) using the injectivity of the Laplace transform.   □

From formulae (6.b) and (6.c)), we now deduce an interesting expression for $E[f(\exp B_t)g(A_t)]$, where $f, g : \mathbb{R}_+ \to \mathbb{R}_+$ are two Borel functions.

Corollary. *We have*

$$E[f(\exp B_t)g(A_t)] = c_t \int_0^\infty dy \int_0^\infty dv \, f(y)g\left(\frac{1}{v}\right) \exp\left(-\frac{v}{2}(1+y^2)\right) \psi_{yv}(t),$$

(6.e)

where

$$c_t = \frac{1}{(2\pi^3 t)^{\frac{1}{2}}} \exp(\pi^2/2t).$$

Proof. By definition of $a_t(x, u)$, we have

$$E[f(\exp B_t)g(A_t)]$$
$$= \frac{1}{\sqrt{2\pi t}} \int_{-\infty}^\infty dx \exp\left(-\frac{x^2}{2t}\right) \int_0^\infty du \, g(u) a_t(x, u) f(e^x)$$
$$= \int_{-\infty}^\infty dx \int_0^\infty \frac{du}{u} \exp\left(-\frac{1}{2u}(1+e^{2x})\right) \theta_{e^x/u}(t) f(e^x) g(u) \quad \text{(from (6.c))}$$
$$= c_t \int_{-\infty}^\infty dx \int_0^\infty \frac{du}{u} \exp\left(-\frac{1}{2u}(1+e^{2x})\right) \frac{e^x}{u} \psi_{e^x/u}(t) f(e^x) g(u) \quad \text{(from(6.b'))}$$
$$= c_t \int_0^\infty dy \int_0^\infty dv \exp\left(-\frac{v}{2}(1+y^2)\right) \psi_{yv}(t) f(y) g\left(\frac{1}{v}\right).$$

□

6.3. From the formulae (6.b) and (6.e), we obtain an 'explicit' expression for the density $\alpha_t(v)$ of A_t, i.e. $P(A_t \in dv) = \alpha_t(v) \, dv$, which is nonetheless complicated.

Indeed, we deduce from (6.e) that, for any Borel function $g : \mathbb{R}_+ \to \mathbb{R}_+$, we have:

$$E[g(A_t)] = c_t \int_0^\infty dv \, g\left(\frac{1}{v}\right) \exp\left(-\frac{v}{2}\right) \alpha_t(v),$$

(6.f)

where

$$\alpha_t(v) = \int_0^\infty dy \psi_{yv}(t) \exp\left(-\frac{vy^2}{2}\right) = \int_0^\infty dx \frac{1}{\sqrt{v}} \psi_{x\sqrt{v}}(t) \exp\left(-\frac{x^2}{2}\right).$$

We then found it interesting to check that, from (6.f), we are able (at least!) to recover the fact that $E[(A_t)^n] < \infty$, for any $n \in \mathbb{N}$. In order to prove this, it suffices to show that $\alpha_t(v) = O(v^k)$, as $v \to 0$, for any $k \in \mathbb{N}$, which, in turn, is implied by the property

$$\psi_r(t) = O(r^k), \quad \text{as} \quad r \to 0, \text{for any } k \in \mathbb{N}. \tag{6.g}$$

Proof of (6.g). The key argument in the proof is the probabilistic representation of $\psi_r(t)$ as

$$\psi_r(t) = \frac{\sqrt{2\pi t}}{2i} E[\exp\left(-r \cosh\left(\sqrt{t}\, G\right)\right)\{\exp \sqrt{t}(1+i\lambda)G - \exp \sqrt{t}\,(1-i\lambda)G\}]$$

where $\lambda = \pi/t$, and G is a standard $\mathcal{N}(0,1)$ gaussian variable. Now, (6.g) follows from the fact that all derivatives of $\psi_r(t)$, with respect to r, when taken at $r = 0$, are equal to 0, which is a consequence of

$$E[(\cosh\left(\sqrt{t}\, G\right))^k \{\exp \sqrt{t}\,(1+i\lambda)G - \exp \sqrt{t}\,(1-i\lambda)G\}] = 0, \tag{6.h}$$

for any $k \in \mathbb{N}$.

In order to prove (6.h), we shall show that, for any $\theta = k\sqrt{t}$, with $k \in \mathbb{Z}$, the quantity

$$h(\theta) \stackrel{\text{def}}{=} E[\exp\left(\theta G\right)\{\exp \sqrt{t}\,(1+i\lambda)G - \exp \sqrt{t}\,(1-i\lambda)G\}]$$

is equal to 0. Indeed, for any $\theta \in \mathbb{R}$, we have

$$h(\theta) = \exp\tfrac{1}{2}(\theta + \sqrt{t}\,(1+i\lambda))^2 - \exp\tfrac{1}{2}(\theta + \sqrt{t}\,(1+i\lambda))^2$$

from which it follows easily, since $\lambda = \pi/t$, that $h(k\sqrt{t}) = 0$, for $k \in \mathbb{Z}$.

7. Some Connections with Hyperbolic Brownian Motion

As announced in Section 1.2, we sketch here how we might deduce the joint law of

$$\left(A_t^{(\nu)} \equiv \int_0^t ds \exp 2(B_s + \nu s), B_t\right)$$

from the knowledge of the semigroup of the hyperbolic Brownian motion.

In order to compute the joint law of $(A_t^{(\nu)}, B_t)$ for fixed t, it is obviously sufficient to compute the conditional law $a_t^\nu(x, u)\, du$ of $A_t^{(\nu)}$, given $B_t = x$.

From the relation (6.a'), we deduce

$$a_t^\nu(x, u) = a_t(x + \nu t, u), \quad du \text{ a.s.} \tag{7.a}$$

In the following, we shall compute (via a Laplace transform) $a_t^\nu(x, u)$ for $\nu = -\frac{1}{2}$, hence for all ν's, thanks to (7.a). We note, for simplicity, $\bar{a}(x, u)$ for $a_t^{-\frac{1}{2}}(x, u)$. Now, the semigroup associated with the hyperbolic Laplacian operator

$$\Delta = \frac{y^2}{2}\left(\frac{\partial^2}{\partial x^2} + \frac{\partial^2}{\partial y^2}\right)$$

on Poincaré's half-plane is known to be [3], [6]:

$$p_t(z; l) = \frac{\sqrt{2}\exp(-\frac{t}{2})}{(2\pi t)^{\frac{3}{2}}}\int_d^\infty r\exp\left(-\frac{r^2}{2t}\right)\frac{dr}{(\cosh r - \cosh d)^{\frac{1}{2}}} \qquad (7.b)$$

where $d = d(z, l)$ is the hyperbolic distance between z and l.

A stochastic differential equation satisfied by the hyperbolic Brownian motion $((X_t, Y_t); t \geq 0)$, starting at $(x, y) \in \mathbb{R} \times \mathbb{R}_+$, with infinitesimal generator Δ, is:

$$X_t = x + \int_0^t Y_s\, dU_s; \quad Y_t = y + \int_0^t Y_s\, dB_s \equiv y\exp\left(B_t - \frac{t}{2}\right), \qquad (7.c)$$

where U and B are two independent real-valued Brownian motions started at 0. Thanks to the independence of U and B, $(X_t, t \geq 0)$ may also be represented as

$$X_t = x + \gamma(A_t^{(-\frac{1}{2})}), \quad \text{where } A_t^{(-\frac{1}{2})} = \int_0^t ds\exp(2B_s - s) \qquad (7.d)$$

and $(\gamma(u), u \geq 0)$ is a one-dimensional Brownian motion starting at 0, which is independent of B. Taking $x = 0$ and $y = 1$, one obtains, for any Borel function $f : \mathbb{R} \times \mathbb{R}_+ \to \mathbb{R}$,

- on the one hand:

$$E[f(X_t, Y_t)] = \int_{\mathbb{R} \times \mathbb{R}_+} p_t(i; (h, k))f(h, k)dh\, dk$$

$$= \int_{\mathbb{R}^2} p_t\left(i; \left(h, \exp\left(x - \frac{t}{2}\right)\right)\right)\exp\left(x - \frac{t}{2}\right)f\left(h, \exp\left(x - \frac{t}{2}\right)\right)dh\, dx;$$

- on the other hand, from (7.c) and (7.d):

$$E[f(X_t, Y_t)] = \frac{1}{\sqrt{2\pi t}}\int_{-\infty}^\infty dx\exp\left(-\frac{x^2}{2t}\right)\ldots$$

$$\ldots\int_0^\infty du\, \bar{a}(x, u)E\left[f\left(\gamma(u), \exp\left(x - \frac{t}{2}\right)\right)\right].$$

From those two expressions for $E[f(X_t, Y_t)]$, we deduce

$$p_t\left(i; \left(h, \exp\left(x - \frac{t}{2}\right)\right)\right) \exp\left(x - \frac{t}{2}\right)$$

$$= \frac{1}{\sqrt{2\pi t}} \exp\left(-\frac{x^2}{2t}\right) \int_{-\infty}^{\infty} du \ \bar{a}(x, u) \frac{1}{\sqrt{2\pi u}} \exp\left(-\frac{h^2}{2u}\right).$$

(7.e)

Changing u to $(1/u)$ in the last integral makes it a Laplace transform (in $h^2/2$) from which $\bar{a}(x, u)$ can—at least in theory—be deduced.

Acknowledgements

The author is grateful to Marc Chesney, Hélyette Geman, Jean-Claude Gruet, Saul Jacka and Nicole El Karoui for their continued interest in this work and many stimulating discussions. J.C. Gruet (personal communication) checked that (7.e) and (6.e) agree, and also developed further computations linking Bessel processes and hyperbolic Brownian motions.

References

1. Bouaziz, L., Briys, E. and Crouhy, M. (1994). The pricing of forward-starting Asian options. *J. Banking and Finance*, **18**, 623–639
2. Bougerol, Ph. (1983). Exemples de théorèmes locaux sur les groupes résolubles. *Ann. Inst. H. Poincaré*, **19**, 369–391
3. Getoor, R.K. (1961). Infinitely divisible probabilities on the hyperbolic plane. *Pacific J. Math.*, **11**, 1287–1308
4. Hartman, P. and Watson, G.S. (1974). 'Normal' distribution functions on spheres and the modified Bessel functions. *Ann. Prob.*, **2**, 593–607
5. Itô, K. and McKean, H.P. (1965). *Diffusion Processes and their Sample Paths*. Springer-Verlag, Berlin
6. Karpalevich, F.I., Tutubalin, V.N. and Shur, M. (1959). Limit theorems for the compositions of distributions in the Lobachevsky plane and space. *Theory Prob. Appl.*, **4**, 399–402
7. Kemna, A.G.Z. and Vorst, A.C.F. (1990). A pricing method for options based on average asset values.*J. Banking and Finance*, **14**, 113–129
8. Lebedev, N.N. (1972). *Special Functions and their Applications*. Dover, New York
9. Pitman, J.D. and Yor, M. (1981). Bessel processes and infinitely divisible laws. In *Stochastic Integrals*, ed. D. Williams. Lecture Notes in Mathematics, **851**, Springer-Verlag, Berlin, 285–370
10. Revuz, D. and Yor, M. (1991). *Continuous Martingales and Brownian Motion*. Springer-Verlag, Berlin
11. Stoyanov, J. (1987). *Counterexamples in Probability*. Wiley, New York
12. Vorst, A.C.F. (1992). Prices and hedge ratios of average exchange rate options. *Intern. Review of Financial Analysis*, **1** (3), 179–193

13. Watson, G.N. (1966). *A Treatise on the Theory of Bessel Functions.* 2nd paper-back edn. Cambridge University Press
14. Williams, D. (1974). Path decomposition and continuity of local time for one-dimensional diffusions I. *Proc. London Math. Soc.*, **28** (3), 738–768
15. Yor, M. (1980). Loi de l'indice du lacet brownien et distribution de Hartman-Watson. *Z. Wahrscheinlichkeits.*, **53**, 71–95
16. Yor, M. (1992). Sur certaines fonctionnelles exponentielles du mouvement brownien réel. *J. Appl. Prob.*, **29**, 202–208. **Paper [1] in this book**

Postscript #2

a) **About Subsection 4.4:** Despite several attempts, finding whether the law of A_t, for fixed t, is determined by its integral moments, has not been resolved.

   J. Stoyanov (2000): *"Krein condition in probabilistic moment problems"*. Bernoulli 6 (5), 939–949

   presents an up to date overall picture of the various sufficient conditions for the determinacy of a distribution (on $\mathbb{R}_+$, say) from its integral moments, and may be helpful to solve the above problem.

b) **About Section 6:** Formulae (6.c) and (6.e) have been useful in a number of studies, e.g.: to obtain the asymptotics in some models of random environments, for instance in the Kawazu-Tanaka paper refered to in article [10] of the present volume.

   For some related discussion of asymptotics of Brownian functionals, see:

   S. Kotani (1996): *"Analytic approach to Yor's formula of exponential additive functionals of Brownian motion"*. In N. Ikeda, S. Watanabe, M. Fukushima, H. Kunita (eds.): *"Itô's stochastic calculus and probability theory"*. Springer, p. 185–196.

c) **About numerical implementations:** In an applied direction, it seems that the presence of the sine function in the integrand in (6.b) makes formula (6.e) difficult to implement for numerical computations related to, say, the price of Asian options. A quite different approach, which yields very interesting lower and upper bounds, has been proposed by:

   L.C.G. Rogers, Z. Shi (1995): *"The value of an Asian option"*. Journal of Applied Probability 32, 1077–1088.

d) **About Section 7:** Starting from formulae (6.e) and (6.c),

   J.C. Gruet (1996): *"Semi-groupe du mouvement Brownien hyperbolique"*. Stochastics and Stochastic Reports 56, 53–61

   obtained new integral representations of the semigroups of hyperbolic Brownian motions. A discussion of the relations between Gruet's representation and more classical formulae, such as (7.b) in the present paper, is made in:

   H. Matsumoto, L. Nguyen, M. Yor (February 2001): *"Subordinators related to the exponential functionals of Brownian bridges and explicit formulae for the semigroups of hyperbolic Brownian motions"*. To appear in the Proceedings of the Siegmundsburg Winter School (February 2000), J. Engelbert, R. Buckdahn eds; Gordon and Breach, publ. (2001).

Some Relations between Bessel Processes, Asian Options and Confluent Hypergeometric Functions[1]

C.R. Acad. Sci., Paris, Sér. I **314** (1992), 471–474

(with Hélyette Geman)

Abstract. A closed formula is obtained for the Laplace transform of moments of certain exponential functionals of Brownian motion with drift, which give the price of some financial options, so-called Asian options. A second equivalent formula is presented, which is the translation, in this context, of some intertwining properties of Bessel processes or confluent hypergeometric functions.

1. The Asian Options Problem: Statement of Results

From the mathematical point of view, the Asian options problem involves giving a closed formula of the greatest possible simplicity for the quantity:

$$C^{(\nu)}(t,k) \overset{\text{def.}}{=} E[(A_t^{(\nu)} - k)^+], \tag{1}$$

where $k \in \mathbb{R}_+$, $\nu \in \mathbb{R}$ and

$$A_t^{(\nu)} = \int_0^t ds \exp 2(B_s + \nu s),$$

in which $(B_s, s \geq 0)$ denotes a real-valued Brownian motion, starting from 0.

The uninitiated reader can gain an idea of the questions of financial mathematics associated with Asian options from [2, 3], in particular.

Here, we give a relatively simple formula for the Laplace transform of $C^{(\nu)}(t,k)$, that is:

$$\int_0^\infty dt\, e^{-\lambda t} C^{(\nu)}(t,k) \equiv \frac{1}{\lambda} E[(A_{T_\lambda}^{(\nu)} - k)^+],$$

for λ sufficiently large, where T_λ denotes an exponential variable with parameter λ, independent of the Brownian motion B.

---

[1] This note was presented by Paul-André Meyer.

Theorem. *Suppose that $n \geq 0$ (not necessarily an integer), and $\lambda > 0$ and set $\mu = \sqrt{2\lambda + \nu^2}$. Suppose that $\lambda > 2n(n+\nu)$, which is equivalent to: $\mu > \nu + 2n$. Then we have, for $x > 0$:*

$$E\left[\left\{\left(A_{T_\lambda}^{(\nu)} - \frac{1}{2x}\right)^+\right\}^n\right]$$

$$= \frac{E[(A_{T_\lambda}^{(\nu)})^n]}{\Gamma((\mu-\nu)/2 - n)} \int_0^x dt \, e^{-t} t^{(\mu-\nu)/2-n-1} \left(1 - \frac{t}{x}\right)^{(\mu+\nu)/2+n}. \quad (2)$$

Moreover, we have

$$E[(A_{T_\lambda}^{(\nu)})^n] = \frac{\Gamma(n+1)\Gamma(((\mu+\nu)/2)+1)\Gamma(((\mu-\nu)/2)-n)}{2^n\Gamma((\mu-\nu)/2)\Gamma(n+((\mu+\nu)/2)+1)}. \quad (3)$$

In the particular case in which n is an integer, this formula simplifies to:

$$E[(A_{T_\lambda}^{(\nu)})^n] = \frac{n!}{\prod_{j=1}^n (\lambda - 2(j^2 + j\nu))}. \quad (4)$$

Decomposing the rational fraction on the right-hand side of (4) into simple elements, we obtain the following closed expression for the moments of $A_t^{(\nu)}$.

Proposition 1. *For all $a \in \mathbb{R}\setminus\{0\}$, $\nu \in \mathbb{R}$ and $n \in \mathbb{N}$, we have:*

$$a^{2n} E\left[\left(\int_0^t du \exp a(B_u + \nu u)\right)^n\right] = n! \sum_{j=0}^n c_j^{(\nu/a)} \exp\left(\left(\frac{a^2 j^2}{2} + aj\nu\right)t\right) \quad (5)$$

where

$$c_j^{(\theta)} = 2^n \prod_{\substack{k \neq j \\ 0 \leq k \leq n}} ((\theta + j)^2 - (\theta + k)^2)^{-1}.$$

In particular, for $\nu = 0$, we have:

$$a^{2n} E\left[\left(\int_0^t du \exp(aB_u)\right)^n\right] = E[P_n(\exp aB_t)] \quad (6)$$

$$\equiv n!\left\{\frac{(-1)^n}{(n!)^2} + 2\sum_{j=1}^n \frac{(-1)^{n-j}}{(n-j)!(n+j)!} \exp\frac{a^2 j^2 t}{2}\right\}$$

where

$$P_n(z) = (-2)^n \left\{ \frac{1}{n!} + 2 \sum_{j=1}^{n} \frac{n!(-z)^j}{(n-j)!(n+j)!} \right\}$$

$$= \frac{(-2)^n}{n!} \{2F(-n, 1, n+1; z) - 1\},$$

in which $F(\alpha, \beta, \gamma; z)$ denotes the hypergeometric function with parameters (α, β, γ).

Remarks. 1. The formula (6) is in agreement with the following result due to Bougerol [1]: for fixed $t \geq 0$, $\sinh(B_t) \overset{\text{dist.}}{=} \gamma_{A_t^{(0)}}$ where $(\gamma_u, u \geq 0)$ is a real-valued Brownian motion independent of $A_t^{(0)}$.

2. Using a different approach to that developed in this paper, one can obtain an expression for the Laplace transform in t of the joint distribution of $(A_t^{(\nu)}, B_t + \nu t)$, in terms of semigroups of Bessel processes, then invert the Laplace transform, but the final expression obtained is complicated (see [8]).

2. Stages in the Proof of the Theorem

2.1. In all that follows, $\nu \geq 0$ is fixed; we simply write A_t to denote $A_t^{(\nu)}$, and $X_t = 2(B_t + \nu t)$, $t \geq 0$. Let us define $\tau_k = \inf\{t : A_t > k\}$. The process $(Y_k = \exp(X_{\tau_k}), k \geq 0)$ is the square of a Bessel process of dimension $\delta_\nu = 2(\nu + 1)$, starting at 1, and we have: $\tau_k = \int_0^k ds/Y_s$.

Let us also define:

$$G_{(n)}^{(\lambda)}(k) = E\left[\int_0^\infty dt\, e^{-\lambda t} \{(A_t - k)^+\}^n \right] \qquad (\lambda, k, n \geq 0),$$

and set $g_{(n)}^{(\lambda)} = G_{(n)}^{(\lambda)}(0)$.

2.2. By application of the strong Markov property, followed or preceded by integration in (dt), the quantity $G_{(n)}^{(\lambda)}(k)$ can be written in the following two equivalent forms

$$G_{(n)}^{(\lambda)}(k) = E[(Y_k)^n \exp(-\lambda \tau_k)] g_{(n)}^{(\lambda)} = \frac{n}{\lambda} \int_k^\infty dv (v-k)^{n-1} E[\exp(-\lambda \tau_v)].$$

2.3. We now let $Q_y^{(\delta)}$ denote the distribution of the square of the Bessel process of dimension $\delta > 0$, starting from y.

Using the local equivalence of the distributions $Q_y^{(\delta_\mu)}$ and $Q_y^{(\delta_\nu)}$ $(y > 0)$ and the closed form of the Girsanov density which relates these probabilities (see, for example, [5, 7]), the two above expressions for $G_{(n)}^{(\lambda)}(k)$ can be rewritten in the form:

$$G_{(n)}^{(\lambda)}(k) = H_\mu\left(n + \frac{\nu - \mu}{2}, k\right) g_{(n)}^{(\lambda)} \tag{7a}$$

$$= \frac{n}{\lambda} \int_k^\infty dv(v - k)^{n-1} H_\mu\left(\frac{\nu - \mu}{2}, v\right), \tag{7b}$$

where we have set: $H_\mu(\alpha, k) = Q_1^{(\delta_\mu)}((Y_k)^\alpha)$, which quantity can be calculated explicitly using the following proposition.

Proposition 2. *The following formulae hold, for* $0 < \gamma < \mu + 1$,

$$\frac{1}{y^\gamma} H_\mu\left(-\gamma, \frac{1}{2y}\right) = Q_y^{(\delta_\mu)}\left(\frac{1}{(Y_{1/2})^\gamma}\right) \tag{8a}$$

$$= \frac{1}{\Gamma(\gamma)} \int_0^1 dt\, e^{-yt} t^{\gamma-1}(1 - t)^{\mu-\gamma}. \tag{8b}$$

Formula (8b) follows easily from the expression for the Laplace transform of the semigroup of the square of a Bessel process, namely:

$$Q_y^{(\delta)}(\exp(-\alpha Y_s)) = \frac{1}{(1 + 2\alpha s)^{\delta/2}} \exp\left(-y\frac{\alpha}{1 + 2\alpha s}\right). \tag{9}$$

2.4. In [10], in addition to the details of the proof of the above theorem, we also present a general Markovian framework in which explicit calculations can be performed; this applies, in particular, to the sawtooth Markov processes studied in [9].

3.   Some Remarkable Identities

3.1. Formula (8b) could equally well have been obtained using the explicit expression for the semigroup of the square of a Bessel process (see, for example, [6], p. 411, Corollary (1.4)). With this approach, one obtains the following formula:

$$\frac{1}{y^\gamma} H_\mu\left(-\gamma, \frac{1}{2y}\right) = \exp(-y)\frac{\Gamma(a)}{\Gamma(b)}\Phi(a, b; y), \tag{10}$$

where $a = -\gamma + 1 + \mu$, $b = 1 + \mu$ and $\Phi(a, b; y)$ denotes the confluent hypergeometric function with parameters a and b.

Using the following classical relations (see Lebedev [4], pp. 266–267)

$$\Phi(a, b; z) = e^z \Phi(b - a, b; -z) \tag{11a}$$

$$\Phi(b - a, b; -z) = \frac{\Gamma(b)}{\Gamma(a)\Gamma(b - a)} \int_0^1 dt\, e^{-zt} t^{(b-a)-1}(1 - t)^{a-1}, \tag{11b}$$

formula (8b), above, is obtained from (10).

3.2. The recursive formula (7b) can be rewritten, after elementary transformations, in the form:

$$x^\alpha H_\mu \left(\alpha, \frac{1}{2x} \right) g_{(n)}^{(\lambda)}$$

$$= \frac{n}{\lambda 2^{n-1}} \int_0^1 dw \, w^{-\alpha-1}(1-w)^{n-1}(xw)^\beta H_\mu \left(\beta; \frac{1}{2wx} \right), \qquad (12)$$

where we have set: $\alpha = n + (\nu - \mu)/2$ and $\beta = (\nu - \mu)/2 \equiv \alpha - n$.

Taking into account formula (8b), the equation (12) is just an analytic translation of the algebraic relationship between beta variables

$$Z_{a,b+c} \stackrel{\text{dist.}}{=} Z_{a,b} Z_{a+b,c}, \qquad (13)$$

for parameter values: $a = -\alpha$, $b = n$, $c = 1 + \mu + \beta$, where the notation $Z_{p,q}$ denotes a beta variable with parameters (p, q), that is, we have:

$$P(Z_{p,q} \in dt) = \frac{t^{p-1}(1-t)^{q-1}}{B(p,q)} dt \quad (0 \leq t \leq 1),$$

where the two variables on the right-hand side of (13) are assumed to be independent. As far as confluent hypergeometric functions are concerned, the relation (12) translates to the identity:

$$\Phi(\alpha, \gamma; z) = \frac{\Gamma(\gamma)}{\Gamma(\beta)\Gamma(\gamma - \beta)} \int_0^1 dt \, t^{\beta-1}(1-t)^{\gamma-\beta-1} \Phi(\alpha, \beta; zt) \ (\gamma > \beta) \quad (14)$$

(see Lebedev [4], p. 278).

3.3. Finally, the different relations (12) and (14) can be understood at the level of the semigroups $(Q_t^{(\delta)}, t \geq 0)$ and $(Q_t^{(\delta')}, t \geq 0)$ of the squares of Bessel processes of dimensions δ and δ', respectively, which are intertwined in the following way:

$$Q_t^\delta M_{\delta'/2,(\delta-\delta')/2} = M_{\delta'/2,(\delta-\delta')/2} Q_t^{\delta'},$$

where $0 < \delta' < \delta$ and $M_{a,b}$ is the multiplicative kernel defined by:

$$M_{a,b} f(x) = E[f(x Z_{a,b})], \quad f \in b(\mathcal{B}(\mathbb{R}_+))$$

(for these intertwining relations, see [9]).

References

1. Bougerol, P. (1983). Exemples de théorèmes locaux sur les groupes résolubles. *Ann. Inst. Henri Poincaré, Probab. Stat.*, **19** (4), 369–391
2. Carverhill, A. and Clewlow, L. (1990). Average-rate options. Risk, **3** (4), 25–29
3. Kemna, A.G.Z. and Vorst, A.C.F. (1990). A pricing method for options based on average asset values. *J. Banking Finance*, **14**, 113–129
4. Lebedev, N.N. (1972). *Special Functions and their Applications.* Dover
5. Pitman, J.W. and Yor, M. (1981). Bessel processes and infinitely divisible laws. In 'Stochastic Integrals' (Lecture Notes in Mathematics, vol. 851) ed. D. Williams, Springer, Berlin
6. Revuz, D. and Yor, M. (1991). *Continuous Martingales and Brownian Motion,* Springer, Berlin
7. Yor, M. (1980). Loi de l'indice du lacet brownien et distribution de Hartman–Watson. *Z. Wahrscheinlichkeit.*, **53**, 71–95
8. Yor, M. (1992). On some exponential functionals of Brownian motion. *Adv. App. Prob.*, **24**, 509–531. **Paper [2] in this volume**
9. Yor, M. (1989). Une extension markovienne de l'algèbre des lois beta–gamma. *C. R. Acad. Sci., Paris, Sér. I*, **308**, 257–260
10. Geman, H. and Yor, M. (1991). Quelques aspects mathématiques du problème des options asiatiques. *See Postscript*

Postscript #3

a) Reference 10. of this note has not been published as such, but its contents are found in Chapter 6 of:

    M. Yor (1992): *"Some aspects of Brownian motion, Part I"*. Lect. in Maths., ETH Zürich, Birkhäuser.

b) Formula (5), in Proposition 1, for the moments of $A_t^{(\nu)}$ goes back at least to Ramakrishnan (1955); see formula (7.44), p. 346 in Chapter 7 of A.T. Bharucha-Reid (1960): *Elements of the Theory of Markov Processes and their Applications.* McGraw-Hill, New-York.

4

The Laws of Exponential Functionals
of Brownian Motion,
Taken at Various Random Times[1]

C.R. Acad. Sci., Paris, Sér. I **314** (1992), 951–956

Abstract. With the help of several different methods, a closed formula is obtained for the laws of the exponential functionals of Brownian motion with drift, taken at certain random times, particularly exponential times, which are assumed to be independent of the Brownian motion.

Abridged Version[2]

Let $(B_t, t \geq 0)$ be a real-valued Brownian motion, starting from 0. For $\nu \in \mathbb{R}$, define:

$$A_t^{(\nu)} = \int_0^t ds \exp 2(B_s + \nu s), \quad t \geq 0.$$

Consider, furthermore, T_λ, an exponential time with parameter λ, independent from B. The main result of this Note is the following:

the distribution of $A_{T_\lambda}^{(\nu)}$ is the same as that of $Z_{1,a}/2Z_b$, where

$$a = \frac{\mu + \nu}{2}, \quad b = \frac{\mu - \nu}{2}, \quad \mu = \sqrt{2\lambda + \nu^2},$$

and $Z_{\alpha,\beta}$ (resp.: Z_γ) denotes a beta variable with parameters (α,β) (resp.: a gamma variable with parameter γ), and these two random variables are assumed to be independent.

This result may be deduced from the closed form of the generalized moments:

$$E[(A_{T_\lambda}^{(\nu)})^\gamma], \text{ which are finite for } \lambda > 2\gamma(\gamma + \nu),$$

and may be expressed in terms of the gamma function, and λ, γ, ν (see [3]).

Conversely, from the identification of the distribution of the law of $A_{T_\lambda}^{(\nu)}$ as that of the beta–gamma ratio presented above, one easily obtains the expression of the generalized moments of $A_{T_\lambda}^{(\nu)}$.

---

[1] This Note was presented by Paul-André Meyer.

[2] This is the *Abridged English Version*, which appeared in the original French paper.

Consider now the case $\nu = 0$, and write, for simplicity, A_t for $A_t^{(0)}$. The knowledge of the distribution of A_{T_λ} enables to recover Bougerol's result (see [1]):

$$\text{for fixed } t \geq 0, \qquad \sinh(B_t) \overset{\text{dist.}}{=} \beta_{A_t},$$

where $(\beta_u, u \geq 0)$ is a real-valued Brownian motion, which is independent of the variable A_t.

In turn, Bougerol's result allows to obtain the laws of A_T for a large class of random times T, assumed to be independent of B.

1.   Some Identities in Distribution

Let $(B_t, t \geq 0)$, be a real-valued Brownian motion, starting at 0. Let $\nu \in \mathbb{R}$ and denote

$$A_t^{(\nu)} = \int_0^t ds \exp 2(B_s + \nu s). \quad (t \geq 0)$$

The explicit expression (1), below, for the moments of the functional $(A_t^{(\nu)})$ taken at an exponential time independent of B is the point of departure for the results of this paper.

Theorem 1. (*See* [3]). *Let* $\gamma \geq 0$, *and* $\lambda > 0$; *set* $\mu = \sqrt{2\lambda + \nu^2}$. *Suppose that* $\lambda > 2\gamma(\gamma + \nu)$, *which is equivalent to:* $\mu > \nu + 2\gamma$. *Then we have*

$$E[(A_{T_\lambda}^{(\nu)})^\gamma] = \frac{\Gamma(1+\gamma)\Gamma(((\mu+\nu)/2)+1)\Gamma(((\mu-\nu)/2)-\gamma)}{2^\gamma\Gamma((\mu-\nu)/2)\Gamma(1+\gamma+((\mu+\nu)/2))} \qquad (1)$$

where T_λ *denotes an exponential time with parameter* λ, *independent of* B.

The identities in distribution (2) and (3), below, can be easily deduced from the identity (1) and, conversely, (1) follows immediately from the identity in distribution (2).

Theorem 2. (*Using the notation of Theorem 1*). *We have the following identity in distribution*

$$A_{T_\lambda}^{(\nu)} \overset{\text{dist.}}{=} \frac{Z_{1,a}}{2Z_b} \overset{\text{dist.}}{=} \frac{1 - U^{1/a}}{2Z_b}, \qquad (2)$$

where $a = (\mu + \nu)/2$, $b = (\mu - \nu)/2$, $Z_{1,a}$ *denotes a beta variable with parameters* $(1, a)$ *and* Z_b *a gamma variable with parameter* b, *that is:*

$$P(Z_{1,a} \in dt) = a(1 - t)^{a-1}dt \quad (0 < t < 1)$$

$$P(Z_b \in dt) = \frac{dt}{\Gamma(b)}t^{b-1}e^{-t} \quad (t > 0),$$

U is a uniform variable on $[0, 1]$. *Finally,* $Z_{1,a}$ *and* Z_b, *respectively,* U *and* Z_b, *are assumed to be independent.*

Corollary 2.1. *Let T be an exponential variable with parameter 1, independent of the Brownian motion B. We have, retaining the notation of Theorem 2, and taking $\lambda = 1$, then respectively: $\nu = 0$, $\nu = 1/2$ and $\nu = -1/2$:*

$$\int_0^{T/2} ds \exp(2B_s) \stackrel{\text{dist.}}{=} \frac{U}{4T} \, ;$$

$$\int_0^T ds \exp(2B_s + s) \stackrel{\text{dist.}}{=} U\sigma; \tag{3}$$

$$\int_0^T ds \exp(2B_s - s) \stackrel{\text{dist.}}{=} \frac{1 - U^2}{2T}$$

where U is a uniform variable on $[0,1]$, $\sigma = \inf\{t : B_t = 1\}$, and the variables on the right-hand sides of the above identities in distribution are assumed to be independent.

Corollary 2.2. *Let $\nu > 0$, then we have:*

$$\int_0^\infty ds \exp 2(B_s - \nu s) \stackrel{\text{dist.}}{=} \frac{1}{2Z_\nu}. \tag{4}$$

The identity in distribution (4) follows easily from (2), in which we let λ tend to 0. This result has already been discussed in [5], and we refer the reader to that article for further details and important references.

Neither is it difficult to derive from the identity (2) an expression, in integral form, for the Laplace transform in t of:

$$E[(A_t^{(\nu)} - k)^+],$$

for all $k \geq 0$, which quantity plays an essential role for the so-called Asian options in financial mathematics (see [3] for a different approach).

Corollary 2.3. *For all $\nu \geq 0$, $\lambda > 2(1 + \nu)$ and $k \geq 0$, we have:*

$$\lambda \int_0^\infty dt \, e^{-\lambda t} E[(A_t^{(\nu)} - k)^+]$$

$$= \frac{\int_0^{1/2k} dt \, e^{-t} t^{((\mu - \nu)/2) - 2} (1 - 2kt)^{((\mu + \nu)/2) + 1}}{(\lambda - 2(1 + \nu))\Gamma(((\mu - \nu)/2) - 1)}.$$

2. The Case $\nu = 0$

We simply let A_t denote $A_t^{(0)}$. The identity in distribution (2) enables us, in particular, to rediscover the following simple result due to Bougerol [1].

Theorem 3. *As before, $(B_t, t \geq 0)$ denotes a real-valued Brownian motion, starting at 0, and $A_t = \int_0^t ds \exp(2B_s)$, $t \geq 0$. Then we have, for any fixed $t \geq 0$:*

$$\sinh(B_t) \stackrel{\text{dist.}}{=} \beta_{A_t}, \tag{5}$$

where, on the right-hand side, $(\beta_u, u \geq 0)$ denotes a real-valued Brownian motion, starting at 0 and independent of B.

Proof.   Let $\theta > 0$ and set $\lambda = \theta^2/2$. To prove (5) it suffices, by the injectivity property of the Laplace transform, to show that:

$$E[|\sinh(B_{T_\lambda})|^\gamma] = E[|\beta_{A_{T_\lambda}}|^\gamma] \tag{6}$$

for all γ sufficiently small.

Now, $|B_{T_\lambda}|$ is an exponential variable with parameter θ; the left-hand side of (6) is therefore equal to:

$$\theta \int_0^\infty dx\, e^{-\theta x} (\sinh x)^\gamma \tag{7}$$

while the right-hand side of (6) is equal to:

$$E(|N|^\gamma) E((A_{T_\lambda})^{\gamma/2}) \tag{8}$$

where N denotes a Gaussian variable with mean 0 and variance 1.

Using, on the one hand, the duplication formula of the gamma function and, on the other hand, formula (1), it is easy to show that expressions (7) and (8) have the common value

$$\frac{1}{2^\gamma} B\left(\frac{\theta - \gamma}{2}, \gamma + 1\right), \quad \text{where} \quad B(\cdot, \cdot) \quad \text{is} \quad \text{Euler's} \quad \text{function.}$$

□

Using Theorem 3, one can exhibit a quite vast family of random times S, independent of B, such that A_S has an interesting distribution. Here are some examples of this.

Theorem 4. *Suppose $\gamma > 0$ and let $T^{(\gamma)}$ denote a random variable with values in $\mathbb{R}_+$, whose distribution is characterized by*

$$E\left[\exp\left(-\frac{\lambda^2}{2} T^{(\gamma)}\right)\right] = \frac{|\Gamma((\gamma + i\lambda)/2)|^2}{(\Gamma(\gamma/2))^2}$$

$$= c_\gamma \int_{-\infty}^\infty \frac{dx}{(\cosh x)^\gamma} \exp(i\lambda x) \tag{9}$$

where

$$c_\gamma = \frac{\Gamma((\gamma + 1)/2)}{\sqrt{\pi}\,\Gamma(\gamma/2)}.$$

We then have:

$$A_{T^{(\gamma)}} \overset{\text{dist.}}{=} \frac{1}{2Z_{\gamma/2}} \tag{10}$$

where, on the left-hand side, the variable $T^{(\gamma)}$ is assumed to be independent of the underlying Brownian motion.

Remark 1. $T^{(\gamma)}$ can be constructed as follows: set $\delta = 2\gamma$; if $(R_\delta(t), t \geq 0)$ denotes the Bessel process of dimension δ, starting at 0, and if we define: $X_{(\gamma)} = \int_0^1 ds R_\delta^2(s)$, we have:

$$E\left[\exp\left(-\frac{x^2}{2} X_{(\gamma)}\right)\right] = \frac{1}{(\cosh x)^\gamma} \quad (x \in \mathbb{R}). \tag{11}$$

Comparing (9) and (11) we see that the distribution of $T^{(\gamma)}$ is characterized by

$$E\left[\exp\left(-\frac{\lambda^2}{2} T^{(\gamma)}\right)\right] = c_\gamma E\left[\sqrt{\frac{2\pi}{X_{(\gamma)}}} \exp\left(-\frac{\lambda^2}{2X_{(\gamma)}}\right)\right]. \tag{12}$$

In the particular case in which $\gamma = 2$, the identity (10) can be rewritten in the following way.

Corollary 4.1. *Let $(R_t, t \geq 0)$, be a Bessel process of dimension 3, starting from 0. We define $T_3 = \inf\{t : R_t = \pi/2\}$. Then we have:*

$$E\left[\exp\left(-\frac{\lambda^2}{2} T_3\right)\right] = \frac{\pi\lambda/2}{\sinh(\pi\lambda/2)} \quad and \quad A_{T_3} \overset{\text{dist.}}{=} \frac{1}{2T}, \tag{13}$$

where T_3 is assumed to be independent of the Brownian motion occurring in the definition of $(A_t, t \geq 0)$ and T is an exponential variable with parameter 1.

Here is another interesting example.

Theorem 5. *We set $V = T_3 + \tilde{T}_3$, where T_3 and $\tilde{T}_3$ are two independent copies of the first passage time to $\pi/2$ for a Bessel process of dimension 3, starting at 0. We then have:*

$$A_V \overset{\text{dist.}}{=} \frac{1}{2\, UT} \tag{14}$$

where V is assumed to be independent of $(A_t, t \geq 0)$ and, on the other hand, U is a uniform variable on $[0, 1]$, independent of the exponential variable T with parameter 1.

3.   Use of Planar Brownian Motion

Suppose $Z_t = X_t + iY_t$, $t \geq 0$ is a complex-valued Brownian motion, starting at 0. P. Lévy has shown that, if $f : \mathbb{C} \to \mathbb{C}$ is a non-constant holomorphic function, then there exists a complex-valued Brownian motion $(\hat{Z}(u), u \geq 0)$, such that:

$$f(Z_t) = \hat{Z} \left(\int_0^t |f'(Z_s)|^2 ds \right), \quad t \geq 0.$$

(For a number of important applications of this result due to Lévy, see B. Davis [2]).

In the particular case in which $f(z) = \exp(z)$, the previous relation becomes:

$$\exp(Z_t) = \hat{Z} \left(\int_0^t \exp(2X_s)ds \right), \quad t \geq 0.$$

The next theorem follows easily from this.

Theorem 6. *Let* $(B_t, t \geq 0)$, *be a real-valued Brownian motion, starting at 0; for* $a > 0$, *we denote* $\sigma_a = \inf\{t : B_t = a\}$.

1. *If* $S = \inf\{t : |Y_t| = \pi/2\}$, *then, we have:*

$$A_S \overset{\text{dist.}}{=} \sigma_1.$$

2. *More generally, let* $\theta \in]-\pi/2, \pi/2]$, *and set* $S_\theta = \inf\{t : Y_t = \theta \text{ or } \theta - \pi\}$. *Then,*

$$A_{S_\theta} \overset{\text{dist.}}{=} \sigma_a, \text{ where } a = |\sin\theta|.$$

Remark 2. It would be interesting to be able to derive some of the results of Section 2, for example, Corollary 4.1 from Lévy's theorem referred to above, but we have not succeeded in doing this.

4.   Comparison with the Results for Fixed Time

Let $t > 0$ and $\nu \in \mathbb{R}$. One of the main results of [6] is the expression for

$$P(A_t^{(\nu)} \in du | B_t + \nu t = x) = a_t(x, u)du,$$

which is given by the formula

$$\frac{1}{\sqrt{2\pi t}} \exp\left(-\frac{x^2}{2t}\right) a_t(x, u) = \frac{1}{u} \exp\left(-\frac{1}{2u}(1 + e^{2x})\right) \theta_r(t), \qquad (15)$$

where $r = e^x/u$, and $(\theta_r(t), t \geq 0)$ is defined by:

$$I_\alpha(r) = \int_0^\infty \exp\left(-\frac{\alpha^2 t}{2}\right) \theta_r(t) dt, \quad \alpha \geq 0, \tag{16}$$

where I_α denotes the modified Bessel function with index α.

The following result issues immediately from (15):

$$P(A_t^{(\nu)} \in du) = du \, \gamma_t^{(\nu)}(u),$$

where:

$$\gamma_t^{(\nu)}(u) = \int_0^\infty dz \, \theta_z(t)(uz)^{\nu-1} \exp\left(-\frac{\nu^2 t}{2}\right) \exp -\frac{1}{2}\left(\frac{1}{u} + uz^2\right). \tag{17}$$

On the other hand, it follows from the identity in distribution (2) that we have:

$$P(A_{T_\lambda}^{(\nu)} \in du) = du \, g_\lambda^{(\nu)}(u),$$

where:

$$g_\lambda^{(\nu)}(u) = \frac{2a}{\Gamma(b)} \int_0^{1/2\,u} dt \, e^{-t} t^b (1 - 2ut)^{a-1}. \tag{18}$$

Now, as a consequence of the very definition of the functions $g_\lambda^{(\nu)}$ and $\gamma_t^{(\nu)}$, we must have the relation:

$$g_\lambda^{(\nu)}(u) = \lambda \int_0^\infty dt \, e^{-\lambda t} \gamma_t^{(\nu)}(u). \tag{19}$$

This is in fact the case, by virtue, for example, of the integral representation of confluent hypergeometric functions using modified Bessel functions (see Lebedev [4], pp. 260–278).

References

1. Bougerol, P. (1983). Exemples de théorèmes locaux sur les groupes résolubles. *Ann. Inst. Henri Poincaré, Probab. Stat.*, **19** (4), 369–391
2. Davis, B. (1979). Brownian motion and analytic functions. *Ann. Probab.*, **7**, 913–932
3. Geman, H. and Yor, M. (1992). Quelques relations entre processus de Bessel, options asiatiques, et fonctions confluentes hypergéométriques. *C. R. Acad. Sci., Paris, Sér. I*, **314**, 471–474. **Paper [3] in this volume**
4. Lebedev, N.N. (1972). *Special Functions and their Applications*. Dover
5. Yor, M. (1992). Sur certaines fonctionnelles exponentielles du mouvement brownien réel. *J. Appl. Probab.*, **29**, 202–208. **Paper [1] in this volume**
6. Yor, M. (1992). On some exponential functionals of Brownian motion. *Adv. App. Prob.*, **24**, 509–531. **Paper [2] in this volume**

Postscript #4

This note aims at giving some examples of closed formulae for the laws of $A_T^{(\nu)}$, when T is assumed independent of $(B_t^{(\nu)} \equiv B_t + \nu t, t \geq 0)$, thus completing the results in paper [2] of this monograph, where T is an exponential variable.

Some analogous studies are made by Matsumoto-Nguyen-Yor (see: Postscript #2) for the exponential functionals $a_t = \int_0^t ds \exp(2b_t(s))$ of Brownian bridges $(b_t(u), 0 \leq u \leq t)$ with arbitrary length $t \geq 0$, when t is replaced by some particular random times.

5

Bessel Processes, Asian Options, and Perpetuities

Mathematical Finance, Vol. **3**, No. 4
(October 1993), 349–375

(with Hélyette Geman)

Abstract. Using Bessel processes, one can solve several open problems involving the integral of an exponential of Brownian motion. This point will be illustrated with three examples. *The first one* is a formula for the Laplace transform of an Asian option which is "out of the money." *The second example* concerns volatility misspecification in portfolio insurance strategies, when the stochastic volatility is represented by the Hull and White model. *The third one* is the valuation of perpetuities or annuities under stochastic interest rates within the Cox–Ingersoll–Ross framework. Moreover, without using time changes or Bessel processes, but only simple probabilistic methods, we obtain further results about Asian options: the computation of the moments of all orders of an arithmetic average of geometric Brownian motion; the property that, in contrast with most of what has been written so far, the Asian option may be more expensive than the standard option (e.g., options on currencies or oil spreads); and a simple, closed-form expression of the Asian option price when the option is "in the money," thereby illuminating the impact on the Asian option price of the revealed underlying asset price as time goes by. This formula has an interesting resemblance with the Black−Scholes formula, even though the comparison cannot be carried too far.

1. Introduction

Bessel processes possess two major features: first, besides the Ornstein–Uhlenbeck processes, they are essentially the only diffusion processes in addition to Brownian motion (with drift), for which a relatively simple expression for the transition probability is known; second, they appear naturally in a number of interesting problems in finance and insurance. For instance, hypergeometric functions (which are related to Bessel processes) are used for the pricing of options on zero coupon bonds in the Cox–Ingersoll–Ross general equilibrium model of interest rates. Another key point is that the standard hypothesis in most financial papers assumes stock price dynamics driven by the exponential of a Brownian motion with drift, which is in turn (the square of) a time-changed Bessel process. Moreover, the class of laws of squares of Bessel processes is stable under the convolution operation, a property which is not shared by the commonly used lognormal models.

We shall illustrate these points with three examples of theoretical and practical importance: Asian options, portfolio insurance, and perpetuities or annuities.

The paper is organized as follows: Section 2 recalls the definition of Bessel processes, their main properties and their relation to exponentials of Brownian motion. A first application is given in Section 3, namely the study of the distribution and the moments of the asset average price, together with a comparison between the Asian and standard option prices and an expression for (the Laplace transform of) the Asian option price. Section 4 explains why Bessel processes can help solve the problem of volatility misspecification in the classical strategies of portfolio insurance which have been extensively implemented over the last decade. Section 5 addresses the pricing of perpetuities and annuities in a stochastic interest rates environment, within the framework of the Cox–Ingersoll–Ross model. Some concluding comments are given in Section 6.

2.   Some Properties of Bessel Processes

In this section, we collect some general results about Bessel processes. Detailed proofs of these results may be found in the references (Shiga and Watanabe 1973, Revuz and Yor 1991, Yor 1992b). For simplicity, we first look at squares of Bessel processes, or "Bessel squared" processes; these we denote by BESQ (a Bessel process BES is the square root of a BESQ).

Definition. For any $\delta \geq 0$, the δ-dimensional Bessel squared process BESQ^δ is a continuous diffusion process ρ_t taking its values in $\mathbb{R}^+$ and satisfying the stochastic differential equation

$$d\rho_t = \delta \, dt + 2\sqrt{\rho_t} \, dW_t, \qquad \rho_0 = a \geq 0,$$

where W_t is a standard one-dimensional Brownian motion.

The real $\nu = \delta/2 - 1$ is called the *index of the process* $BESQ^\delta$.

As a consequence of the stochastic differential equation satisfied by $(\rho_t, \, t \geq 0)$, the process

$$\left\{ f(\rho_t) - \int_0^t ds(L^\nu f)(\rho_s), t \geq 0 \right\}$$

is a local martingale as soon as f belongs to $C^2((0, +\infty))$, where

$$L^\nu f(\rho) = 2\rho \frac{d^2 f}{d\rho^2} + \delta \frac{df}{d\rho},$$

and L^ν is the (martingale) infinitesimal generator associated with BESQ^δ.

We denote by Q_a^δ the distribution of the process BESQ^δ starting at $a \geq 0$; this distribution is defined on the canonical space of continuous functions $C(\mathbb{R}^+, \mathbb{R}^+)$ on which we consider the coordinate process $(R_t, t \geq 0)$ which is defined by $R_t(f) = f(t)$ for every $f \in C(\mathbb{R}^+, \mathbb{R}^+)$ and the σ-field $\mathscr{G} = \sigma\{R_t, t \geq 0\}$.

The following important property of the Bessel squared processes was obtained by Shiga and Watanabe (1973).

Proposition 2.1. *For every* $\delta, \delta', x, x' \geq 0, Q_x^\delta \otimes Q_{x'}^{\delta'} = Q_{x+x'}^{\delta+\delta'}$, *where* $P \otimes Q$ *denotes the distribution of the process* $(X_t + Y_t, t \geq 0)$, *for* X_t *and* Y_t, *two independent processes with respective distributions* P *and* Q.

This property permits, whether δ is an integer or not, the reduction of a number of problems involving BESQ$^\delta$ to the case $\delta = 1$, where BESQ^1 is precisely the square of one-dimensional Brownian motion.

Bessel (squared) processes are, by definition, Markov processes. Their transition functions have the following form.

Proposition 2.2. *For* $\delta > 0$, *the semigroup of* BESQ$^\delta$ *has a density in* y *equal to*

$$q_t^\delta(x, y) = \frac{1}{2t} \left(\frac{y}{x}\right)^{\nu/2} \exp\left(-\left(\frac{x+y}{2t}\right)\right) I_\nu\left(\frac{\sqrt{xy}}{t}\right) \quad (t > 0, x, y > 0),$$

(2.1)

where I_ν *is the Bessel function with index* $\nu \equiv \delta/2 - 1$.

The density of the semigroup of the Bessel process BES$^\delta$ can be obtained from (2.1) by a straightforward change of variable and is found equal, for $\delta > 0$, to

$$p_t^\delta(x, y) = \frac{1}{t} \left(\frac{y}{x}\right)^\nu y \exp\left[-\frac{x^2 + y^2}{2t}\right] I_\nu\left(\frac{xy}{t}\right) \quad (t > 0, x, y > 0). \quad (2.2)$$

A key result for our applications in the following sections is due to Williams (1974).[1]

Proposition 2.3. *The exponential of Brownian motion with drift is a time-changed Bessel process; more specifically,*

$$\exp(W(t) + \nu t) = R^{(\nu)}\left(\int_0^t \exp 2(W(s) + \nu s) ds\right) \quad (t \geq 0),$$

where $(R^{(\nu)}(u), u \geq 0)$ *is a Bessel process with index* ν.

Random time changes are of constant use in the study of one-dimensional diffusions and are likely to be more widely used in finance. The reader interested in the issue of "stochastic clocks" can find a number of important examples in Itô and McKean (1965). Moreover, one can observe that

---

[1] An earlier reference is the paper of Lamperti (1972) quoted in paper [8] in this book; see also Postscript #1.

this issue already arises in adjustments made by practitioners to account for different temporal rates of trading. The "business time scale" and the "transaction clock" which are often introduced in the study of intraday prices involve indeed stochastic time scale changes.

We can give two extensions of Proposition 2.3.

1. A similar result holds more generally for $(\exp(aW(t) + \nu t), t \geq 0)$. Thanks to the scaling property of Brownian motion, we can write

$$\exp(aW(t) + \nu t) = \exp\left(\hat{W}(ta^2) + \frac{\nu}{a^2}ta^2\right),$$

(with $\hat{W}$ another Brownian motion, and corresponding $\hat{R}$ from Proposition 2.3)

$$= \hat{R}^{(\nu/a^2)}\left(\int_0^{ta^2} \exp 2\left(\hat{W}(s) + \frac{\nu s}{a^2}\right)ds\right)$$

$$= \hat{R}^{(\nu/a^2)}\left(a^2 \int_0^t \exp 2(\hat{W}(a^2 u) + \nu u)du\right),$$

so that

$$\exp(aW(t) + \nu t) = \hat{R}^{(\nu/a^2)}\left(a^2 \int_0^t \exp 2\left(aW(u) + \nu u\right)du\right). \tag{2.3}$$

2. The exponential of Brownian motion with drift is a time-changed Bessel squared process.

This property is straightforward since, from (2.3), we deduce

$$\exp(aW(t) + \nu t) = \exp 2\left(\frac{a}{2}W(t) + \frac{\nu t}{2}\right)$$

$$= \left[\hat{R}^{(2\nu/a^2)}\left(\frac{a^2}{4}\int_0^t \exp 2\left(\frac{a}{2}W(s) + \frac{\nu}{2}s\right)ds\right)\right]^2.$$

More generally, it is possible to write the exponential of a Brownian motion with drift as any given power of a Bessel process, up to a time change. However, the additivity property of squares of Bessel processes makes it most efficient to use the power 2.

Proposition 2.4. *Denote by $P_a^{(\nu)}$ the law on $C(\mathbb{R}^+, \mathbb{R}^+)$ of the Bessel process $R^{(\nu)}$ starting at a, by $(R_t, t \geq 0)$ the canonical process on $C(\mathbb{R}^+, \mathbb{R}^+)$, and by $\mathscr{G}_t = \sigma(R_s, s \leq t)$ the canonical filtration. Then, the following relationship holds (from Girsanov's theorem):*

$$P_a^{(\nu)}\big|_{\mathscr{G}_t} = \left(\frac{R_t}{a}\right)^\nu \exp\left(-\frac{\nu^2}{2}\int_0^t \frac{ds}{R_s^2}\right) \cdot P_a^{(0)}\big|_{\mathscr{G}_t}.$$

We shall denote by $E^{(\nu)}$ the expectation relative to $P_1^{(\nu)}$.

As a consequence of Proposition 2.4, we obtain

Lemma 2.1. *For any $\nu \geq 0, \lambda \geq 0$, define $\mu = \sqrt{2\lambda + \nu^2}$ and $\tau_t = \int_0^t ds/(R_s)^2$. Then for every $t \geq 0$ and $\alpha \in \mathbb{R}$ we have*

$$E^{(\nu)}[(R_t)^\alpha \exp(-\lambda \tau_t)] = E^{(0)}\left[(R_t)^{\alpha+\nu} \exp\left(-\frac{\mu^2}{2}\tau_t\right)\right].$$

Proposition 2.5. *Using the same notations as in the definition at the beginning of this section, the following relationship holds for any $\theta > 0, a \geq 0, \mu > \theta - 1$:*

$$E_a^{(\mu)}\left(\frac{1}{(R_t)^{2\theta}}\right) = \frac{1}{\Gamma(\theta)} \int_0^1 h^{\theta-1}(1-h)^{\mu-\theta}\left(\frac{1}{2t}\right)^\theta \exp\left(\frac{-ah}{2t}\right) dh.$$

where Γ denotes the gamma function.

The proof of Proposition 2.5 may be found for instance in Yor (1992a, Chapter 6).

Proposition 2.6. *(Yor 1980). For every $\nu \in \mathbb{R}, a > 0, x > 0, t > 0$,*

$$E_a^{(0)}\left[\exp\left(\frac{-\nu^2}{2}\int_0^t \frac{ds}{R_s^2}\right) \middle| R_t = x\right] = \frac{I_{|\nu|}}{I_0}\left(\frac{ax}{t}\right).$$

where, on the right-hand side, $(f/g)(y)$ stands for $f(y)/g(y)$.

3. Asian Options Revisited

3.1. The General Problem

Even though a number of results have been established on the subject of Asian options (defined below), so far no one has produced an exact formula for their values. Nevertheless, Asian options are popular in the financial community, because they might be superior to standard options for several reasons. First, as we will see later, they are often cheaper than the equivalent classical European option (or are thought to be so even when it is not the case). Second, the fact that the option is based on an average price is an attractive feature for thinly traded assets and commodities, e.g., gold or crude oil, where price manipulations near the option expiration date are possible; this is the case for index options in newly opened markets during the famous "triple-witching" hour. Today, average options represent an enormous percentage of options on oil; some are written directly on oil prices, others on spreads between two types of oil. In the same way, Asian options on average exchange rates are extremely popular because they are less expensive for corporations which need to hedge a series of risky positions on the foreign exchange incurring at a steady rate over a period of time. Some options on domestic interest rates (and options on interest rate swaps) exhibit this Asian feature when the base rate is an arithmetic average of spot rates.

In this paper, we shall address only the "fixed-strike" case, because the "floating-strike" case, where the payoff depends on the relative values of the arithmetic average and the underlying price at maturity, is less widely used in practice. The exact pricing of fixed-strike Asian options is clearly a difficult exercise because the distribution of an arithmetic average of stock prices is unknown (and in particular *not* lognormal) when prices themselves are lognormally distributed. The studies on this problem fall roughly into three groups.

The *first approach* is numerical: Kemna and Vorst (1990) use Monte Carlo simulations with the corresponding geometric average option as a control variable; Carverhill and Clewlow (1990) use the fast Fourier transform to calculate the density of the sum of random variables as the convolution of individual densities and then numerically integrate the payoff function against the density. Clearly, these methods, besides being computationally expensive, give no information on the hedging portfolio.

A *second approach*, used by Ruttiens (1990) and Vorst (1992), consists of modifying the geometric average option price. Although this method reduces the amount of calculation, it may lead to significantly underpriced call options and has the same shortcoming as the numerical one in terms of dynamic hedging. Bouaziz, Bryis, and Crouhy (1991) provide an upper bound of the error involved in their approximation of the law of the average.

A *third approach*, proposed by Lévy (1992) and adopted by a number of practitioners, assumes that the distribution of an arithmetic average is well approximated—at least in some markets—by the lognormal distribution, and, consequently, the problem is reduced to determining the necessary parameters. This task is less complicated since the first two moments of the arithmetic average are relatively simple to derive. In what follows, we shall give a closed-form expression for all moments of the arithmetic average.

3.2. The Mathematical Setting

We take as given a complete probability space $(\Omega, \mathscr{F}, P)$ with a filtration $(\mathscr{F}_t)_{0 \leq t \leq \infty}$, which is right-continuous and such that $\mathscr{F}_0$ contains all the P-null sets of $\mathscr{F}$. Following the seminal papers by Harrison and Kreps (1979) and Harrison and Pliska (1981), we assume the existence of a "risk-neutral" probability measure Q (equivalent to P) under which the underlying stock price dynamics are driven by

$$dS(t) = rS(t)\,dt + \sigma S(t)\,dW_t.$$

Here, r is the instantaneous risk-free rate and σ is the volatility. We suppose that $(\mathscr{F}_t)$ is the natural filtration of the Brownian motion W_t; for simplicity, we assume in this section that both r and σ are constant over $[t_0, T]$, the time interval over which the average value of the stock is calculated, where T is the maturity of the option. In what follows, we will use interchangeably the notation W_t or $W(t)$.

Consider the process $(A(x), x \geq 0)$, where

$$A(x) = \frac{1}{x - t_0} \int_{t_0}^{x} S(u) \, du.$$

The payoff of the Asian option at maturity T is then

$$\max[(A(T) - k), 0] = (A(T) - k)^+$$

where k is the fixed-strike price of the option. By arbitrage arguments and because we assumed the interest rate to be constant over the lifetime of the option, the value at time t of the Asian call option is

$$C_{t,T}(k) = e^{-r(T-t)} E_Q[(A(T) - k)^+ | \mathscr{F}_t].$$

We start with three simplifying steps.

Step 1.

$$C_{t,T}(k) = e^{-r(T-t)} E_Q[(A(T) - k)^+ | \mathscr{F}_t] = \frac{e^{-r(T-t)}}{T - t_0} \bar{C}_{t,T}(k')$$

where

$$\bar{C}_{t,T}(k') \stackrel{\text{def}}{=} E_Q[(\bar{A}(T) - k')^+ | \mathscr{F}_t], \bar{A}(T) = \int_{t_0}^{T} S(u) \, du, k' = k(T - t_0).$$

Step 2. We now show how to reduce the computation of this conditional expectation to that of the deterministic function

$$\zeta^{a,b}(s, \gamma) \stackrel{\text{def}}{=} E\left[\left(\int_0^s \exp(aW(u) + bu) \, du - \gamma\right)^+\right],$$

where $b \in \mathbb{R}$, $a, s, \gamma \in \mathbb{R}$.

Consider the case $t_0 \leq t < T$. (The case $t_0 > t$ of the "forward-start" Asian option is solved in Appendix A). We have

$$E_Q\left[(\bar{A}(T) - k')^+ | \mathscr{F}_t\right] = E_Q\left[\left\{S(t) \int_0^{T-t} \exp(\hat{y}(s)) \, ds - (k' - \bar{A}(t))\right\}^+ \middle| \mathscr{F}_t\right].$$

using

$$\bar{A}(T) = \bar{A}(t) + S(t) \int_0^{T-t} \exp(\hat{y}(s)) \, ds,$$

with $\hat{y}(s) \stackrel{\text{def}}{=} \sigma \hat{W}(s) + (r - \sigma^2/2)s$, where $\hat{W}(s) = W(s+t) - W(t), s \geq 0$, is a new Brownian motion independent of $\mathscr{F}_t$. Hence,

$$E_Q[(\bar{A}(T) - k')^+ | \mathscr{F}_t] = S(t) E_Q \left[\left(\int_0^{T-t} \exp(\hat{y}(s)) \, ds - k'' \right)^+ \Big| \mathscr{F}_t \right],$$

where $k'' = (k' - \bar{A}(t))/S(t)$. Consequently, $\bar{C}_{t,T}(k') = S(t) \zeta^{\sigma,b}(T - t, k'')$, with $b = r - \sigma^2/2$.

From now on, we will drop the index Q in the expectation symbol.

Step 3. We show finally that the scaling property of Brownian motion allows to replace σ by 2 in the expression

$$\zeta^{\sigma,b}(s, \gamma) = E \left[\left(\int_0^s \exp(\sigma W(u) + bu) \, du - \gamma \right)^+ \right].$$

Using the change of variable $u = 4\nu/\sigma^2$ and the scaling property of Brownian motion, we obtain

$$\zeta^{\sigma,b}(s, \gamma) = \frac{4}{\sigma^2} \zeta^{2,4b/\sigma^2} \left(\frac{s\sigma^2}{4}, \frac{\gamma\sigma^2}{4} \right).$$

Finally, defining

$$C^{(\nu)}(x, q) = E \left[\left(\int_0^x \exp\{2(W(u) + \nu u)\} \, du - q \right)^+ \right],$$

we can write

$$\zeta^{\sigma,b}(s, \gamma) = \frac{4}{\sigma^2} C^{(2b/\sigma^2)} \left(\frac{s\sigma^2}{4}, \frac{\sigma^2\gamma}{4} \right).$$

To summarize steps 1–3, we just expressed the desired Asian option price as

$$C_{t,T}(k) = \frac{e^{-r(T-t)}}{T - t_0} \left(\frac{4S(t)}{\sigma^2} \right) C^{(\nu)}(h, q) \tag{3.1}$$

where

$$\nu = \frac{2r}{\sigma^2} - 1; \ h = \frac{\sigma^2}{4}(T - t); \ q = \frac{\sigma^2}{4S(t)} \left\{ k(T - t_0) - \int_{t_0}^t du \, S(u) \right\}.$$

We must remark at this point that q may be negative if the observed values of the underlying asset S over the time interval $[t_0, t]$ are big enough (this means we already know at time t that the Asian option will be "in the money" at maturity T and that the put option is worthless).

3.3. First Results

We now compare the Asian option price to the standard option price. As shown above, if very high values of $S(t)$ have been observed on the interval $[t_0, t]$, the Asian option is already very valuable a long time before maturity. So, for our comparison to be of interest, we assume that $t_0 = t = 0$; that is, no price of the underlying asset relevant to the average has been revealed when we compare the Asian option and the standard option prices. The quantity k represents the exercise price and is positive.

Proposition 3.1. *The Asian call option price is smaller than the standard option price when the underlying asset is a domestic stock; i.e., $\nu > 0$.*

Proof. This result can be proved in the general case, without the discretization procedure used by Kemna and Vorst (1990), who were themselves looking at a stock Asian option. Moreover, as observed by Turnbull and Wakeman (1991), Kemna and Vorst did not emphasize the fact that the property does not hold when the time to maturity of the call option is shorter than the length of the averaging period.

From the equality

$$\frac{1}{T} \int_0^T \exp 2(W_s + \nu s)\, ds - k = \frac{1}{T} \int_0^T [\exp 2(W_s + \nu s) - k]ds$$

we deduce, thanks to Jensen's inequality,

$$E\left\{ \left(\frac{1}{T} \int_0^T \exp 2(W_s + \nu s)\, ds - k \right)^+ \right\} \leq \frac{1}{T} \int_0^T E\{ (\exp 2(W_s + \nu s) - k)^+ \}ds.$$

The convexity of the exponential function implies that $\{\exp 2(W_s + \nu s), s \geq 0\}$ is a submartingale as long as $\nu = 2r/\sigma^2 - 1$ is positive: this is usually the case for the values of r and σ observed in practice when the underlying asset is a domestic stock, say $r = 0, 1; \sigma = 0, 2; \nu = 4$. Consequently, $\{(\exp 2(W_s + \nu s) - k)^+, s \geq 0\}$ is also a submartingale and

$$E[(\exp 2(W_s + \nu s) - k)^+] \leq E[(\exp 2(W_T + \nu T) - k)^+].$$

$\square$

We can give a more precise result and show that the crucial element in determining whether this property holds is the sign of $\nu + 1 = 2r/\sigma^2$, i.e., the sign of the drift r of the underlying asset. For currency Asian options, for instance, the dynamics of the underlying asset are

$$\frac{dS}{S} = (r - y)\, dt + \sigma\, dW_t, \qquad \text{where } y \text{ is the foreign spot rate.}$$

Consequently, the "risk-neutral" drift $r - y$ may be negative. The same observation holds for Asian options on oil spreads.

A more general version of Proposition 3.1 is the following.

Proposition 3.2. (i) *If $\nu + 1 \geq 0$, the Asian option price is smaller than the standard option price for any exercise price k.*

(ii) *If $\nu + 1 < 0$, the Asian option price is greater than the standard option price for all values of the exercise price k in a neighborhood of 0.*

Proof.   Using Itô's lemma, we can write

$$\exp[2(W(s) + \nu s)]$$
$$= 1 + 2 \int_0^s \exp[2(W(u) + \nu u)] \, dW_u + 2 \int_0^s (\nu + 1) \exp[2(W(u) + \nu u)] \, du.$$

Consequently, $\{\exp[2(W(s) + \nu s)], s \geq 0\}$ is a submartingale if $\nu + 1 \geq 0$ and is a supermartingale if $\nu + 1 \leq 0$. In particular, it is a martingale for $\nu = -1$.

(i) When $\nu + 1 \geq 0$, the submartingale property remains satisfied for $(\exp 2(W(s) + \nu s) - k)^+$; hence, the result of Proposition 3.1 holds for any value of k.

(ii) When $\nu + 1 < 0$, $\{\exp 2(W(s) + \nu s), s \geq 0\}$ is a strict supermartingale and if we choose $k = 0$, i.e., if we look at a zero-exercise price option,[2] we obtain the inequality

$$E \left[\frac{1}{T} \int_0^T \exp 2(W(s) + \nu s) \, ds \right] > E(\exp[2(W(T) + \nu T)]). \qquad (3.2)$$

We could have avoided the supermartingale argument because we can make a simple explicit computation of both sides of (3.2). The left-hand side is equal to $(1 - \exp(-b))/b$, where $b = -2(\nu + 1)T$, and, keeping in mind the equality $E[\exp(2W(s))] = \exp(2s)$, the right-hand side is equal to $\exp(-b)$.

Now, the elementary inequality $(1 - e^{-b})/b > e^{-b}$ concludes this second proof of (3.2). Furthermore, for any k in the interval $[0, \{(1 - \exp(-b))/b\} - \exp(-b)]$, the following inequality holds:

$$E \left[\left(\frac{1}{T} \int_0^T \exp 2(W(s) + \nu s) \, ds - k \right)^+ \right] \geq E[(\exp 2(W(T) + \nu T) - k)^+].$$

$$(3.3)$$

---

[2] Zepos were first introduced by Garman and Kohlhagen (1983).

To prove (3.3), we first use Jensen's inequality to obtain

$$E\left[\left(\frac{1}{T}\int_0^T \exp 2(W(s)+\nu s)\,ds - k\right)^+\right]$$

$$\geq \left(E\left[\left(\frac{1}{T}\int_0^T \exp 2(W(s)+\nu s)\,ds\right)\right] - k\right)^+.$$

From the simple computation relative to the case $k = 0$, the right-hand side is equal to $((1-\exp(-b))/b-k)^+$, which, for k in the prescribed interval, is greater than (or equal to)

$$\exp(-b) = E[\exp 2(W(T)+\nu T)] \geq E[(\exp 2(W(T)+\nu T)-k)^+],$$

the last inequality being obvious.

This result does not hold for all k, at least for $\nu \in [-2,-1]$ and probably for every $\nu < -1$. $\qquad\square$

If we set

$$A_t^{(\nu)} = \int_0^t \exp[2(W_s+\nu s)]\,ds,$$

the distribution of the pair $(A_t^{(\nu)}, W_t + \nu t)$ was obtained by Yor (1992b). First, it is useful to observe that, due to Girsanov's theorem relating the laws of $(W_s + \nu s; s \leq t)$ and $(W_s; s \leq t)$, the quantity $P(A_t^{(\nu)} \in du \,|\, W_t + \nu t = x)$ depends on t, x, u, but not on ν. Consequently, we can take $\nu = 0$, in which case

$$P(A_t^{(0)} \in du, W_t \in dx) = \frac{1}{u}\exp\left[-\frac{1}{2u}(1+e^{2x})\right]\phi_{e^x/u}(t)\,du\,dx, \qquad (3.4)$$

where

$$\phi_\alpha(t) = \frac{\alpha}{(2\pi^3 t)^{1/2}}\exp\left(\frac{\pi^2}{2t}\right)\psi_\alpha(t)$$

and

$$\psi_\alpha(t) = \int_0^\infty dy\, e^{-y^2/2t} e^{-\alpha(\mathrm{ch}\,y)}\,\mathrm{sh}\,y\,\sin\frac{\pi y}{t}.$$

(We use the French notation ch, sh, and th for, respectively, the hyperbolic cosine, sine, and tangent functions.)

Given the complexity of the exact joint distribution of the pair {average, underlying asset}, we shall look at the moments of the average $A_t^{(\nu)}$, which give some partial information about its law. (However, the valuation of an Asian option having a payoff $\max(A_T^{(\nu)} - S_T, 0)$ would require the use of the joint distribution of $(A_T^{(\nu)}; S_T)$.)

The first two moments of $A_t^{(\nu)}$ are easily computed by integration. Higher order moments have been computed by Dufresne (1989, 1990) using Itô's lemma, time reversal, and a recurrence argument. A different method consists of introducing the Laplace transform in time; see Geman and Yor (1992). Using either procedure, one obtains the formula[1]

$$
E\left[\left(\int_0^t \exp \lambda(W_s + \nu s)\, ds\right)^n\right] = \frac{n!}{\lambda^{2n}} \left\{\sum_{j=0}^n d_j^{(\nu/\lambda)} \exp\left[\left(\frac{\lambda^2 j^2}{2} + \lambda j \nu\right)t\right]\right\},
$$

where

$$
d_j^{(\beta)} = 2^n \prod_{\substack{i \neq j \\ 0 \le i \le n}} [(\beta + j)^2 - (\beta + i)^2]^{-1}.
$$

For $\lambda = 2$, this expression gives the different moments of $A_t^{(\nu)}$. In particular, for $n = 1$, we obtain the first-order moment of $A_t^{(\nu)}$:

$$
E(A_t^{(\nu)}) = \frac{1}{2(\nu + 1)}[\exp(2(\nu + 1)t) - 1], \tag{3.5}
$$

an expression which could have been obtained by a direct calculation, interchanging the order of integration and expectation, thanks to Fubini's theorem.

The first- and second-order moments would give the parameters of the distribution of $A_t^{(\nu)}$ if it were lognormal, which is the approximation made by some practitioners, although there is clearly no mathematical justification for this assumption. A possibly related fact is that practitioners are very interested in knowing the extent to which a distribution can be characterized by its moments. A relevant mathematical result is the following (see for instance Feller 1964, vol. 2, p. 224).

Carleman's Criterion. *If a random variable X satisfies $\sum_{n \in \mathbb{N}} (m_{2n})^{-1/2n} = \infty$, where $m_{2n} = E(X^{2n})$, then, its distribution is determined by its moments.*

The criterion does not apply to lognormal distributions nor to the average of geometric Brownian motion.

---

[1] See Postscript #3.

Proposition 3.3. *For any $\nu \in \mathbb{R}$ and $t > 0$, the quantity*

$$M_t^{(\nu)} \stackrel{\text{def}}{=} \sum_{n=1}^{\infty} E[(A_t^{(\nu)})^{2n}]^{-1/2n}$$

is finite.

Proof.   (a) We first reduce the proof to the case $\nu = 0$. If $\nu \geq 0$, $A_t^{(\nu)} \geq A_t^{(0)}$ so $M_t^{(\nu)} \leq M_t^{(0)}$. On the other hand, if $\nu \leq 0$,

$$A_t^{(\nu)} \geq e^{2\nu t} A_t^{(0)} \quad \text{and} \quad M_t^{(\nu)} \leq e^{2\nu t} M_t^{(0)}.$$

(b) To show that $M_t^{(0)} < \infty$, we use Bougerol's identity (1983). For any fixed $t > 0$, $\operatorname{sh} B_t \stackrel{(\text{law})}{=} \gamma_{A_t}$, where we denote $A_t^{(0)}$ by A_t and γ represents a Brownian motion independent of A_t. From the scaling property of Brownian motion, we deduce

$$E[(\operatorname{sh} B_t)^{4n}] = E[A_t^{2n}]E[\gamma_1^{4n}].$$

Hence, in order to show that $M_t^{(0)} < \infty$, it remains to prove that the series with general term $\theta_n = a_n/b_n$, where $a_n = [E(\gamma_1^{4n})]^{1/2n}$ and $b_n = [E(\operatorname{sh}(B_t))^{4n}]^{1/2n}$, is convergent. We now observe that

$$E[(\gamma_1)^{2p}] = (2p/\sqrt{\pi})\Gamma(p + 1/2);$$

hence, using Stirling's formula, we obtain

$$a_n \underset{n\to\infty}{\simeq} (4n/e).$$

On the other hand, we claim that

$$E[(\operatorname{sh}(B_t))^{4n}]^{1/2n} \underset{n\to\infty}{\simeq} E\left[\left(\frac{1}{2}\exp(|B_t|)\right)^{4n}\right]^{1/2n}. \tag{3.6}$$

Assuming (3.6) for now, we deduce that, for n large enough, there exists $\varepsilon \in {]}0, 1{[}$ such that

$$b_n \geq (1 - \varepsilon)\left(E\left[\left(\frac{1}{2}\exp(B_t)\right)^{4n}\right]\right)^{1/2n} \geq \frac{1 - \varepsilon}{4}\exp(4nt).$$

Consequently, $\theta_n = a_n/b_n \leq Cn\exp(-4nt)$ and the series $(\theta_n)_{n\geq 0}$ converges. The equivalence (3.6) is derived from the following chain of equivalences:

$$\left(E\left[(\operatorname{sh}(B_t))^{4n}\right]\right)^{1/2n} \underset{n\to\infty}{\simeq} \left(E\left[\left(\operatorname{sh}(B_t)\right)^{4n} 1_{(|B_t|\geq 1)}\right]\right)^{1/2n}$$

$$\underset{n\to\infty}{\simeq} \left(E\left[\left(\frac{1}{2}\exp(|B_t|)\right)^{4n} 1_{(|B_t|\geq 1)}\right]\right)^{1/2n}$$

$$\underset{n\to\infty}{\simeq} \left(E\left[\left(\frac{1}{2}\exp(|B_t|)\right)^{4n}\right]\right)^{1/2n}.$$

$\square$

3.4.   Asian Option Valuation

Let us return to (3.1), obtained at the end of Section **3.2**. What remains to be computed is the quantity $C^{(\nu)}(h,q) \overset{\text{def}}{=} E[(A_h^{(\nu)} - q)^+]$, where $A_h^{(\nu)} = \int_0^h \exp[2(W_s + \nu s)]\,ds$.

(a) When $q \leq 0$, the calculation is straightforward:

$$C^{(\nu)}(h,q) = E(A_h^{(\nu)}) - q.$$

Using (3.5) we obtain

$$C^{(\nu)}(h,q) = \left\{\frac{1}{2(\nu+1)}[\exp(2(\nu+1)h) - 1]\right\} - q.$$

This quantity, plugged into (3.1) with the corresponding values of $\nu, h,$ and q gives a closed-form expression of the Asian option price, namely,

$$C_{t,T}(k) = S(t)\left(\frac{1 - e^{-r(T-t)}}{r(T-t_0)}\right) - e^{-(T-t)}\left(k - \frac{1}{T-t_0}\int_{t_0}^t S(u)\,du\right). \quad (3.7)$$

This expression deserves some comment.

(1) It has an interesting resemblance to the Black and Scholes (1973) formula (it is easy to show that the coefficient of $S(t)$ is smaller than 1), but, as shown in the following remarks, the comparison should not be carried too far.

(2) The volatility σ does not appear explicitly in the call price, but it is carried implicitly in $S(t)$ and in the integral $\int_{t_0}^t S(u)\,du$. This appears clearly in a direct and simple proof of (3.7), using the property that the discounted underlying asset price is a Q-martingale; moreover, the result easily extends to stochastic interest rates (see Geman 1989), since it indeed involves the forward price of the underlying asset.

(3) It is obviously satisfactory to observe that $C_{t,T}(k)$ is simultaneously and separately increasing in the integral $\int_{t_0}^{t} S(u) \, du$ and in $S(t)$.

(4) The "delta" and the "gamma" of the Asian call can be immediately derived from (3.7). It is worth observing that gamma is zero while delta is not constant, which is an uncommon situation. This property reflects the fact that the hedging strategy consists in selling every day (or continuously) the same fraction of the underlying asset.

For $q > 0$, we do not have such a simple reduction of $C^{(\nu)}(h, q)$, but we are able to provide an expression of its Laplace transform with respect to the variable h, which mirrors the fact that $A^{(\nu)}$ taken at an independent exponential time has a remarkable form (see Yor 1992c).

First, we recall from (3.5) that

$$c^{(\nu)}(h) \overset{\text{def}}{=} C^{(\nu)}(h, 0) = \frac{1}{2(\nu+1)} \{\exp[2(\nu+1)h] - 1\}.$$

Then, using Proposition 2.3 we can represent the process $X_x = \exp(W_x + \nu x), x \geq 0$ as

$$X_x = R^{(\nu)}(A_x^{(\nu)}),$$

where $(R^{(\nu)}(u), u \geq 0)$ is a Bessel process with index ν, starting at $R^{(\nu)}(0) = 1$. Moreover, if $\tau_u^{(\nu)} = \inf\{s; A_s^{(\nu)} > u\}$, then

$$\tau_u^{(\nu)} = \int_0^u \frac{ds}{(R^{(\nu)}(s))^2} \quad \text{and} \quad R^{(\nu)}(u) = X(\tau_u^{(\nu)}).$$

Now, we write

$$A_h^{(\nu)} = A_{\tau_q^{(\nu)}}^{(\nu)} + \int_{\tau_q^{(\nu)}}^{h} \exp 2(W_s + \nu s) \, ds,$$

on the set $(A_h^{(\nu)} \geq q)$. Since $A_{\tau_q^{(\nu)}}^{(\nu)} = q$ and denoting $R_q^{(\nu)} = [X(\tau_q^{(\nu)})]$, in agreement with (2.3), then the strong Markov property and the independence of the increments of Brownian motion imply that, for all $h \geq 0$ and $q \geq 0$,

$$E\left[(A_h^{(\nu)} - q)^+ \mid \mathscr{F}_{\tau_q^{(\nu)}}\right] = (R_q^{(\nu)})^2 c^{\nu}[(h - \tau_q^{(\nu)})^+].$$

Consequently,

$$C^{(\nu)}(h, q) = E[(R_q^{(\nu)})^2 c^{\nu}((h - \tau_q^{(\nu)})^+)]. \tag{3.8}$$

From (3.8) we can derive the Laplace transform in h of $C^{(\nu)}(h, q)$. For $\lambda > \delta$,

$$\int_0^\infty e^{-\lambda h} C^{(\nu)}(h, q)\, dh = D_\nu(q, \lambda) \frac{1}{\lambda(\lambda - \delta)},$$

where

$$D_\nu(q, \lambda) \stackrel{\text{def}}{=} E[(R_q^{(\nu)})^2 \exp(-\lambda \tau_q^{(\nu)})].$$

Using Proposition 2.6, i.e.,

$$E[\exp(-\lambda \tau_q^{(\nu)}) \mid R_q^{(\nu)} = s] = \frac{I_\mu(s/q)}{I_\nu(s/q)},$$

where $\mu = \sqrt{2\lambda + \nu^2}$, we can write

$$D_\nu(q, \lambda) = \int_0^\infty p_q^{(\nu)}(1, s) s^2 \frac{I_\mu(s/q)}{I_\nu(s/q)} ds.$$

Here $p_q^{(\nu)}(1, s)$ denotes the density at time q of the Bessel semigroup, with index ν and starting position 1 at time 0; this density is given by Proposition 2.2 (where we used p^δ instead of $p^{(\nu)}$, with $\delta = 2(1 + \nu)$):

$$p_q^{(\nu)}(1, s) = \frac{s^{\nu+1}}{q} \exp\left(-\frac{1}{2q}(1 + s^2)\right) I_\nu\left(\frac{s}{q}\right).$$

Consequently,

$$D_\nu(q, \lambda) = \int_0^\infty \frac{1}{q} \exp\left(-\frac{1}{2q}(1 + s^2)\right) s^{3+\nu} I_\mu\left(\frac{s}{q}\right) ds$$

$$\tag{3.9}$$

$$= \frac{1}{\Gamma((\mu - \nu)/2 - 1)} \int_0^{1/2q} e^{-x} x^{(\mu-\nu)/2 - 2} (1 - 2qx)^{(\mu+\nu)/2 + 2} dx,$$

where $\mu = \sqrt{2\lambda + \nu^2}$ and where the integrals in (3.10) are two representations of confluent hypergeometric functions (see Lebedev 1972, p. 278, Exercise 12, and p. 266, (9.11.1) and (9.11.2)).

Finally, the Laplace transform of $C^{(\nu)}(h, q)$ with respect to the variable h can be written as

$$\int_0^\infty dh\, e^{-\lambda h} C^{(\nu)}(h, q) = \frac{\int_0^{1/2q} dx\, e^{-x} x^{(\mu-\nu)/2 - 2}(1 - 2qx)^{(\mu+\nu)/2 + 1}}{\lambda(\lambda - 2 - 2\nu)\Gamma((\mu - \nu)/2 - 1)}, \tag{3.10}$$

where Γ denotes the gamma function.

The inversion of this Laplace transform for a fixed h would provide the quantity $C^{(\nu)}(h, q)$ and, hence, the Asian option price thanks to (3.1). This inversion is not easy, since the parameter λ also appears in the gamma function. There is now off-the-shelf software for inverting Laplace transforms, which may be useful for a numerical solution of this problem.

4. Quadratic-Variation-Based Strategies

In this section, we consider the issue of portfolio insurance based on option replication when the model of the underlying asset does not have a constant volatility over the investment period, which is the case in practice.

The most straightforward method of insuring a portfolio of risky assets is to purchase a put option on the portfolio, with a strike price equal to some desired minimum value for the portfolio. Alternatively, a dynamic portfolio of stock and cash can be created to replicate the stock-plus-put portfolio. However, the construction of such synthetic portfolio insurance will depend crucially on a correct estimation of the volatility of the portfolio being insured.

Rendleman and O'Brien (1990), among others, have shown that misspecification of the volatility can cause the outcome of a synthetic portfolio insurance strategy to deviate significantly from its target. Assuming that the Black–Scholes option pricing model (particularly the hypothesis of a constant volatility over the lifetime of the option) is appropriate for determining the value of the put option, a very simple argument—which is the core of the Black–Scholes formula—gives the number of shares of the stock one must hold and the amount one must invest in a riskless asset (e.g., T-bills). These numbers depend on time and on the market value of the risky asset, and, consequently, portfolio adjustments have to be made on a continuous basis; however, the relative amounts in stocks and T-bills also depend on the volatility (through the coefficients d_1 and d_2 in the Black–Scholes formula). Therefore, errors in volatility estimation will result in an incorrect mix during the life of the insurance program. This issue has usually been addressed by estimating the deviation from the target at the time horizon H (using, for instance, simulations of volatility, as in Hill et al. 1988).

Bick (1991) offers a new approach, which is to build "quadratic-variation-based" dynamic strategies: instead of facing an unwanted outcome of the portfolio insurance strategy at the time H chosen as the horizon, he proposes to stop the strategy at the first time τ_a such that

$$\int_0^{\tau_a} \sigma^2(s)\,ds = a, \tag{4.1}$$

where $\sqrt{a}$ is the volatility parameter chosen at inception of the portfolio strategy at time $t = 0$, and $\sigma(t)$, the true volatility of the risky asset, is not necessarily constant or deterministic but only $\mathscr{F}_t$-measurable. It is easy to see that, assuming σ is continuous and $\mathscr{F}_t$-adapted, the stopping time

$$\tau_a = \inf\left\{ u \geq 0;\ \int_0^u \sigma^2(s)\,ds = a \right\}$$

is the first instant at which the option replication is correct and, consequently, the portfolio value is equal to the desired target.

Pursuing this approach, we can calculate the distribution of τ_a in the Hull and White (1987) framework. Denoting by S the risky asset and by $y = \sigma^2$

the instantaneous variance, we suppose that their dynamics are driven by the following stochastic differential equations:

$$\frac{dS}{S} = \mu_1 \, dt + \sigma \, dW_1(t); \frac{dy}{y} = \mu_2 \, dt + \xi \, dW_2(t), \quad \text{where } y(0) = y_0,$$

and $d\langle W_1, W_2 \rangle_t = \rho_t \, dt$; i.e., W_1 and W_2 are two $\mathscr{F}_t$-Brownian motions with a deterministic bracket.

It will be convenient to use the notation $\mu_2' = \mu_2/2$ and $\eta = \xi/2$. Hence, we have

$$\sigma(t) = \sqrt{y_0} \exp\{(\eta) \, W_2(t) + (\mu_2' - \eta^2)t\}.$$

Making the change of variables $t = u/\eta^2$ and using the scaling property of Brownian motion, we can write

$$\sigma(t) = \sigma\left(\frac{u}{\eta^2}\right) = \sigma_0 \exp\left[\bar{W}_2(u) + (\mu_2' - \eta^2)\frac{u}{\eta^2}\right],$$

where $\bar{W}_2$ is another Brownian motion. Recognizing the exponential of a Brownian motion with drift, we use again its Bessel process representation:

$$\sigma\left(\frac{u}{\eta^2}\right) = R_{\sigma_0}^{(\nu)}\left(\int_0^u ds \, \sigma^2\left(\frac{s}{\eta^2}\right)\right), \tag{4.2}$$

where $(R_{\sigma_0}^{(\nu)}(t), t \geq 0)$ is a Bessel process starting at σ_0, with index $\nu = \mu_2'/\eta^2 - 1$.

We now wish to identify the stopping time τ_a in (4.1). To simplify computations, we first assume $\eta = 1$. Making in (4.1) the random time change $s = \tau_b$, we obtain $\tau_a = \int_0^a db/\sigma^2(\tau_b)$. Now, replacing u by τ_b in (4.2) gives $\sigma(\tau_b) = R_{\sigma_0}^{(\nu)}(b)$. Hence,

$$\tau_a = \int_0^a \frac{db}{(R_{\sigma_0}^{(\nu)}(b))^2}.$$

Similar computations for a general η lead to

$$\tau_a = \frac{1}{\eta^2} \int_0^{2a\eta^2} \frac{db}{(R_{\sigma_0}^{(\nu)}(b))^2}.$$

The probability density $f_a(x)$ of τ_a does not have a simple expression (see Yor 1980, Sections 4–6), but its Laplace transform is relatively simple, being precisely

$$\int_0^\infty f_a(x)e^{-\lambda x}dx = E_{\sigma_0}^{(\nu)}[\exp(-\lambda\tau_a)] = E_{\sigma_0}^{(\nu)}\left[\exp -\frac{\lambda}{\eta^2}\int_0^{2a\eta^2}\frac{db}{(R_{\sigma_0}^{(\nu)}(b))^2}\right]$$

$$= E_{\sigma_0}^{(\mu)}\left(\frac{\sigma_0}{R_{2a\eta^2}}\right)^{2\kappa},$$

where Lemma 2.1 was applied twice; here, $\mu = (2\lambda/\eta^2 + \nu^2)^{1/2}$ and $\kappa = (\mu - \nu)/2$.

Using Proposition 2.5 we can transform the latter expectation into

$$\int_0^{+\infty} f_a(x)e^{-\lambda x}\,dx = \frac{1}{\Gamma(\kappa)} \int_0^1 \left\{\exp\left(-\frac{u\sigma_0^2}{2a\eta^2}\right)\right\} \left(\frac{u\sigma_0^2}{2a\eta^2}\right)^\kappa (1-u)^{\mu-\kappa}du.$$

Therefore, we have replaced uncertainty about the portfolio outcome by uncertainty about the variable time horizon t of the strategy; however, we know the distribution of this stopping time. For practical purposes, this approach to volatility misspecification in classical portfolio insurance strategies removes the following unpleasant feature of these strategies, which is that the level of investment of the portfolio in the risky asset just prior to the fixed horizon H is either 0% or 100%, depending on whether the value of the portfolio at that time is below or above the desired floor. An investor wishing to resume this strategy would see his portfolio reallocated to a drastically different asset mix, whereas his risk aversion has not changed from one day to the next. In the "stopping time investment strategies" that we propose, the investor, after having obtained the desired terminal wealth, can choose any type of new strategy. Longstaff (1990) had already raised the issue of European options with "extendible maturities"; in our approach, however, the date at which the target portfolio is exactly obtained—i.e., the date t at which $\sqrt{a}$ was the correct volatility to use in the Black and Scholes formula—may be earlier than the horizon H at which the target would have been reached in the situation of constant volatility. In this case, this date t would be the right time to resume the portfolio insurance.

In the same spirit, many fund managers have been particularly interested since the market crash of 1987 in hedging specifically against sudden market drops. Some financial institutions today offer index options (assuming the manager holds a well-diversified portfolio) which have a fixed exercise price and a stochastic maturity date as defined above. If there is a market collapse, the high volatility in (4.1) implies that the put option is rapidly exercised (i.e., τ_a is small), protecting the fund value; otherwise, the option keeps its time value. These options are less expensive than American options (which also have a random maturity date but of another nature) and still allow one to avoid the drawbacks of synthetic portfolio insurance strategies during market crashes.

5. Valuation of a Perpetuity in a Stochastic Interest Rates Environment

In a recent paper, Dufresne (1990) was able to establish the distribution of a perpetuity making a continuous payment of one unit over time, when this perpetuity is expressed as

$$Z = \int_0^\infty dt \exp -(\sigma W_t + \nu t)$$

where ν and σ are positive.

In fact, Dufresne showed that Z is distributed as the reciprocal of a gamma variable. Taking into account, in the actuarial approach, the stochastic nature of interest rates is certainly an interesting step. However, this approach has two limitations: first, the result does not hold (see Section 3) when the upper bound of the integral in Z is a fixed number T (which would better express the finite nature of human life): second, and more troublesome, is that the discount factor used by Dufresne, $\exp -(\sigma W_t + \nu t)$, is radically different from the one which would be derived from any of the models presented in the abundant financial literature on interest rates. This may be related to Dufresne's discount factor being obtained as the limit of discrete discounting over a number of time intervals going to infinity. We are going to take these two points into account in our approach, but at the cost of losing the knowledge of the exact distribution of the random variable Z.

We take the classical setting in which the short-term rate is the state variable of the interest rates movements and is driven by the following dynamics:

$$dr_t = (\delta + \beta r_t)\, dt + \sigma r_t^\gamma\, dW_t, \quad r(0) = r_0. \tag{5.1}$$

Here, W_t denotes a standard Brownian motion, δ, β, σ, and γ are constants and $1/2 \le \gamma \le 1$ or $\gamma = 0$ (see, for instance, Geman and Portait 1989).

We will first make an important observation: if $\gamma \in\,]0, 1/2[$, (4.1) may not have a unique solution; at least in the case $\beta = \delta = 0$, it is well known that this equation does not have a unique solution. This famous result is due to Girsanov (see, for instance, Rogers and Williams 1987, vol. 2, pp. 175–176). When $\gamma = 1/2$, it can be shown that if $2\delta \ge \sigma^2$, (5.1) has a unique solution which does not hit 0.

The situation $\gamma = 1/2, \beta < 0$ corresponds to the well-known Cox–Ingersoll–Ross (1985) model of interest rates. The C.I.R. term structure model is derived in a general equilibrium setting; not only does it avoid arbitrage opportunities, but it also provides the market price of interest rate risk as part of the equilibrium. In the C.I.R. case, the coefficient β in (4.1) under the risk-neutral probability incorporates the market price of interest rate risk. Within this setting, we are going to price perpetuities and annuities.

Denoting $\beta = -2\theta$, we deduce from (5.1) that for any $\lambda \in \mathbb{R}_+^*$,

$$r(\lambda t) = r_0 + (\delta\lambda)t + \sigma \int_0^{\lambda t} \sqrt{r(s)}\, dW_s - 2\theta\lambda \int_0^t r(\lambda s)\, ds.$$

Using the scaling property of Brownian motion, we can write

$$r(\lambda t) = r_0 + (\delta\lambda)t + \sigma \int_0^t \sqrt{r(\lambda s)}\sqrt{\lambda}\; d\bar{W}_s - 2\theta\lambda \int_0^t r(\lambda s)\, ds \tag{5.2}$$

where $\bar{W}_s$ represents another Brownian motion. Denoting $\bar{\delta} = \delta\lambda, \bar{\theta} = \theta\lambda$, and $\bar{r}(t) = r(\lambda t)$ and choosing $\lambda = 4/\sigma^2$, we observe that when $\bar{\delta}$ is an integer, $\bar{r}(t)$ is the square of the norm of an Ornstein–Uhenbeck process $(\bar{x}(t), t \geq 0)$ with dimension $\bar{\delta}$ and parameter $\bar{\theta}$, i.e.,

$$\bar{x}(t) = x_0 + \bar{z}_t - \bar{\theta} \int_0^t \bar{x}(s) \, ds, \tag{5.3}$$

where $(\bar{z}_t, t \geq 0)$ is a $\bar{\delta}$-dimensional Brownian motion and $x_0 \in \mathbb{R}^{\bar{\delta}}$. A proof of this result is obtained by using Itô's lemma for $\rho(t) = |\bar{x}(t)|^2$:

$$\rho(t) = |x_0|^2 + 2 \int_0^t \sqrt{\rho(s)} \, d\bar{W}_s - 2\bar{\theta} \int_0^t \rho(s) \, ds + \bar{\delta}t,$$

where $d\bar{W}_s = (\bar{x}(s) \cdot d\bar{z}(s))/|\bar{x}(s)|$ (the numerator being an inner product in $\mathbb{R}^{\bar{\delta}}$).

We have thus recovered (5.1) since $\sigma\sqrt{\lambda} = 2$. When $\bar{\delta} = 1$, (5.3) involves the one-dimensional Brownian motion and $\bar{r}(t)$ is the square of an Ornstein–Uhlenbeck process. This (one-dimensional) Ornstein–Uhlenbeck process has become familiar in finance since it was introduced by Vasicek (1977) when defining the short-term interest rate dynamics by the particular form of (5.1):

$$dr = a(b - r)dt + \sigma dW_t, \quad \text{with} \quad a, b, \text{ and } \sigma \text{ constant.}$$

We have thus exhibited a relationship between two very popular models of the term structure dynamics. The Gaussian feature of the Vasicek model has the drawback of possibly negative interest rates (but would lead to a mathematically simple valuation of annuities and perpetuities). Consequently, we are going to study the pricing of perpetuities in the Cox–Ingersoll–Ross framework.

Prior to that, we want to show how the valuation of zero coupon bonds in the C.I.R. model can be obtained within our methodology. The price at time 0 of a zero coupon bond maturing at time t is

$$B(0, t) = E_Q \left[\exp \left(- \int_0^t r(s) \, ds \right) \right].$$

Making the change of variable $s = \lambda u = 4u/\sigma^2$ in the integral, we obtain

$$B(0, t) = E_Q \left[\exp \left(- \int_0^{4t/\sigma^2} \bar{r}(u) \frac{4}{\sigma^2} du \right) \right].$$

where $\bar{r}(u)$ is the square of the norm of $\bar{x}(u)$, a $\bar{\delta}$-dimensional Ornstein–Uhlenbeck process with parameter $\bar{\theta} = 4\theta/\sigma^2$. Girsanov's relationship between the law $P_{\bar{\theta}}$ of the norm of $\bar{x}(u)$ and the law P of the $\bar{\delta}$-dimensional Bessel process $R(u)$, assuming they both start at $\sqrt{r_0}$ at time 0, is

$$P_{\bar{\theta}}|_{\mathscr{F}_u} = \exp\left\{ -\frac{\bar{\theta}}{2}(R_u^2 - \bar{\delta}u - r_0) - \frac{\bar{\theta}^2}{2}\int_0^u (R(x))^2 \, dx \right\} \cdot P\,|_{\mathscr{F}_u}.$$

Moreover, formula (2.k) in Pitman and Yor (1982) establishes that for any a and b in $\mathbb{R}^+$,

$$E\left[\exp\left(-\frac{a}{2}R_t^2 - \frac{b^2}{2}\int_0^t R_s^2 \, ds\right)\right]$$

$$= \left(\mathrm{ch}(bt) + \frac{a}{b}\mathrm{sh}(bt)\right)^{-\bar{\delta}/2} \exp\left(\frac{-r_0 b}{2}\frac{a/b + \mathrm{th}(bt)}{1 + (a/b)\mathrm{th}(bt)}\right),$$

where again the $\bar{\delta}$-dimensional Bessel process $R(t)$ originates at $\sqrt{r_0}$. Combining these two results, we obtain for the bond price at time 0 that

$$B(0,t) \equiv UVW,$$

with

$$U = \exp\left(\frac{2\theta}{\sigma^2}\left(r_0 + \frac{16\delta t}{\sigma^4}\right)\right),$$

$$V = \left\{\mathrm{ch}\left(\frac{16t}{\sigma^3}\left(\frac{1}{2} + \frac{\theta^2}{\sigma^2}\right)^{1/2}\right) + \frac{\theta}{\sigma}\left(\frac{1}{2} + \frac{\theta^2}{\sigma^2}\right)^{-1/2}\mathrm{sh}\left(\frac{16t}{\sigma^3}\left(\frac{1}{2} + \frac{\theta^2}{\sigma^2}\right)^{1/2}\right)\right\}^{2\delta/\sigma^2},$$

$$W = \exp\left(-\frac{2r_0}{\sigma}\left(\frac{1}{2} + \frac{\theta^2}{\sigma^2}\right)^{1/2}\right)\cdots$$

$$\cdots\left\{\frac{(\theta/\sigma)(1/2 + \theta^2/\sigma^2)^{-1/2} + \mathrm{th}((16t/\sigma^3)(1/2 + \theta^2/\sigma^2)^{1/2})}{1 + (\theta/\sigma)(1/2 + \theta^2/\sigma^2)^{-1/2}\ \mathrm{th}((16t/\sigma^3)(1/2 + \theta^2/\sigma^2)^{1/2})}\right\}.$$

A perpetuity making a continuous payment of one unit over time has the form

$$Z = \int_0^\infty \exp\left(-\int_0^t r(s)\,ds\right)\,dt$$

$$= \lambda\int_0^\infty du\,\exp\left(-\lambda\int_0^u r(\lambda s)\,ds\right), \quad \text{for any } \lambda > 0.$$

To gain "simplicity," we first look at the case $\beta = 0, \gamma = 1/2$. Defining $b^2 = 8/\sigma^2 = 2\lambda$, we see that Z takes the form

$$Z = \lambda \int_0^\infty \exp\left(-\frac{b^2}{2}\int_0^t ds\, [R(s)]^2\right) dt \stackrel{\text{def}}{=} \lambda Z_1. \tag{5.4}$$

where $(R(s), s \geq 0)$ denotes a Bessel process with dimension $\delta\lambda = 2(\nu + 1)$, starting from $\sqrt{r_0}$.

We are interested in computing the first-order moment of Z since its distribution is not simple. The details of this computation, which use roughly the same tools as the ones used in the previous sections, are completed in Appendix B and the result is given in (B.5):

$$E(Z) = \frac{\sqrt{2}}{\sigma}\int_0^1 \frac{1}{\sqrt{h}}(1 - h)^{(\nu-1)/2}\exp\left(\frac{-br_0}{2}\sqrt{h}\right) dh,$$

where $b = 2\sqrt{2}/\sigma, \nu = 4\delta/\sigma^2 - 1$, and δ and σ are the parameters of the short-term dynamics in the Cox–Ingersoll–Ross model defined in (5.1) by $\gamma = 1/2$ (and $\beta = 0$ for now).

Denoting by Z_T an annuity which is continuously paid between dates 0 and T, all the previous results may be slightly modified to show that the financial value of this annuity at time 0 is

$$E[Z_T] = \frac{\sqrt{2}}{\sigma}\int_0^{\text{th}(bT)}(1 - u^2)^{(\nu-1)/2}\exp\left(-\frac{r_0 bu}{2}\right) du$$

with the same notation as above. (We remark that, taking the derivative of this expression with respect to T, we recover the bond price $B(0, T)$ calculated above, for $\theta = 0$.)

As a consequence, we obtain

$$E[Z] = \frac{\sqrt{2}}{\sigma}\int_0^1 \frac{dh}{\sqrt{h}}(1 - h)^{(\nu-1)/2}\exp\left(-\left(\frac{br_0}{2}\right)\sqrt{h}\right)$$

$$= \frac{\sqrt{2}}{\sigma}\left\{\sum_{n=0}^\infty \frac{(-1)^n}{n!}\left(\frac{br_0}{2}\right)^n \int_0^1 dh\, (1 - h)^{(\nu-1)/2}h^{(n-1)/2}\right\}$$

$$= \frac{\sqrt{2}}{\sigma}\left\{\sum_{n=0}^\infty \left(-\frac{br_0}{2}\right)^n \frac{1}{n!}\beta\left(\frac{\nu+1}{2}, \frac{n+1}{2}\right)\right\},$$

where β is the (Euler) beta function.

Now, from the expression of β in terms of Γ and the duplication formula for Γ, i.e.,

$$\Gamma(n + 1) = \frac{2^n}{\sqrt{\pi}}\Gamma\left(\frac{n}{2} + 1\right)\Gamma\left(\frac{n+1}{2}\right),$$

we deduce

$$E[Z] = \frac{\sqrt{2}}{\sigma} \left\{ \sum_{n=0}^{\infty} \left(-\frac{br_0}{2} \right)^n \frac{\Gamma((\nu+1)/2)\sqrt{\pi}}{2^n \Gamma(n/2+1)\Gamma((n+\nu)/2+1)} \right\}.$$

In the general case of the Cox–Ingersoll–Ross model, $\beta = -2\theta$, with $\theta > 0$ and the same method gives

$$E(Z) = \lambda \exp\left(\frac{\theta \lambda r_0}{2} \right)$$

$$\int_0^{+\infty} dt \exp\left(\frac{\theta \delta \lambda^2}{2} t \right) (\mathrm{ch}\,\omega t + \underline{\theta}\,\mathrm{sh}\,\omega t)^{-\delta\lambda/2} \exp\left\{ \frac{-r_0\,\omega}{2} \left(\frac{\theta + \mathrm{th}\,(\omega t)}{1 + \underline{\theta}\,\mathrm{th}\,(\omega t)} \right) \right\}$$

where $\omega = \sqrt{2\lambda + \theta^2\lambda^2}$, $\lambda = 4/\sigma^2$, $\underline{\theta} = \theta\lambda/\omega$, and the parameters β, δ, and σ define the short-term rate dynamics in (5.1) with $\gamma = 1/2$. Introducing the change of variables $u = \mathrm{th}(\omega t)$, we can write

$$E(Z) = \lambda \exp\left(\frac{\theta \lambda r_0}{2} \right) \int_0^1 \frac{du}{\omega(1-u^2)} \left(\frac{1+u}{1-u} \right)^{(\delta\lambda/4\omega)\underline{\theta}} \cdots$$

$$\cdots \left(\frac{1}{\sqrt{1-u^2}} + \underline{\theta} \frac{u}{\sqrt{1-u^2}} \right)^{-\delta\lambda/2} \exp\left\{ \frac{-r_0\omega}{2} \left(\frac{\theta+u}{1+\underline{\theta}u} \right) \right\}.$$

After simple computations, this can be rearranged as

$$E(Z) = \frac{\lambda}{\omega} \exp\left(\frac{\theta \lambda r_0}{2} \right) \int_0^1 (1+u)^p (1-u)^q (1+\underline{\theta}u)^{-\delta\lambda/2} \cdots$$

$$\cdots \exp\left\{ \frac{-r_0\,\omega}{2} \left(\frac{\theta+u}{1+\underline{\theta}u} \right) \right\} du$$

where

$$p = \frac{\theta \delta \lambda^2}{4\omega} + \frac{\delta\lambda}{4} - 1, \quad q = \frac{\delta\lambda}{4} - \frac{\theta\delta\lambda^2}{4\omega} - 1, \quad \omega = (2\lambda + \theta^2\lambda^2)^{1/2}, \quad \lambda = \frac{4}{\sigma^2}.$$

The same type of computational developments as in the case $\beta = \theta = 0$ could be made from here.

6. Conclusion

We have seen how Bessel processes can be used to bring two kinds of answers to the problem of Asian options pricing, one through the calculation of all moments of the average of the underlying asset price, the other through the expression (of the Laplace transform) of the Asian option price. In the portfolio insurance framework, Bessel processes provide alternative answers to the issue of deviating from the target at the time horizon in portfolio insurance strategies when the volatility of the risky asset is stochastic. In the same manner, Bessel processes appear to be the appropriate tool to price perpetuities and annuities in a stochastic interest rates environment when the dynamics of the short-term rate are defined by a particular case of the Cox–Ingersoll–Ross model. These processes also appear very promising for other models of the term structure dynamics.

Appendix A.
The "Forward-Start" Asian Option

A.1.   Reduction of the Integral

We now show that for $t < t_0 < T$, it is possible to reduce the valuation problem to the case $t = 0$:

$$E[(\bar{A}(T) - k)^+ | \mathcal{F}_t] = E\{[E(\bar{A}(T) - k)^+ | \mathcal{F}_{t_0}] | \mathcal{F}_t\} = E[\phi(S(t_0)) | \mathcal{F}_t],$$

where

$$\phi(s) = E\left[\left(s \int_0^T du \exp(y(u) - y(t_0)) - k\right)^+\right],$$

since $S(u) = \exp[y(u)] = S(t_0) \exp[y(u) - y(t_0)]$ and $(y(t), t \geq 0)$ has independent increments. We observe that

$$\phi(s) = sE\left[\left(\int_0^T du \exp(y(u) - y(t_0)) - \frac{k}{s}\right)^+\right] \stackrel{\text{def}}{=} s\,\bar{C}_{t_0,T}(k/s).$$

Now, conditioning with respect to $\mathcal{F}_t$, we obtain

$$E[\phi(S(t_0)) | \mathcal{F}_t] = E\left[S(t_0) \bar{C}_{t_0,T}\left(\frac{k}{S(t_0)}\right) | \mathcal{F}_t\right] \stackrel{\text{def}}{=} \chi(S(t)),$$

where

$$\chi(s) = sE\left[\frac{S(t_0)}{S(t)}\,\bar{C}_{t_0,\,T}\left(\frac{k}{s}\frac{S(t)}{S(t_0)}\right)\right]$$

$$= sE\left[\exp\{y(t_0 - t)\}\bar{C}_{t_0,\,T}\left(\frac{k}{s}\exp{-y(t_0 - t)}\right)\right].$$

The gain is that the last expression is no longer a conditional expectation but simply an expectation.

A.2.   The Use of This Instrument

When, in 1989, Dow Chemical made a tender offer on Marion Laboratory, the shareholders of the latter company received in exchange for an old share a "contingent value right" attached to a new share. This contingent value right (C.V.R.), which was traded on the American Stock Exchange between 1989 and 1991, was exactly an Asian put option, since the payoff in September 1991 was \$45.77 – \$A where \$A was the average value of the stock during the last 60 days. The same contingent value right was used in the tender offer of the French company Rhône-Poulenc on the American firm Rorer. The Rhône-Poulenc-Rorer contingent value rights are currently traded on the American Stock Exchange; their expiration date is July 1993.

Appendix B.
Valuation of a Perpetuity
in the C.I.R Framework

We start from (5.4):

$$Z = \lambda Z_1 = \lambda \int_0^\infty \exp\left(-\frac{b^2}{2}\int_0^t [R(s)]^2 ds\right)dt.$$

It will be very helpful to use the relation

$$qR_\nu^{1/q}(t) = R_{\nu q}\left(\int_0^t ds R_\nu^{-2/p}(s)\right), \tag{B.1}$$

where $(R_m(t), t \geq 0)$ denotes a Bessel process with index m, and $1/p + 1/q = 1$ (see Revuz and Yor 1991, Chap. 11). [*Note:* In this Appendix, we use the notation R_ν instead of $R_{(\nu)}$, as in the rest of the paper, because here we deal constantly with powers of R_ν, and $R_\nu^\alpha(t)$ denotes the quantity $R_\nu(t)$ raised to the power α.]

If we take $p = -1$, we obtain $q = 1/2$. Denoting

$$A_u = \inf\left\{t : \int_0^t ds R_\nu^2(s) > u\right\},$$

we deduce from (B.1), where we replace t by A_u, that

$$qR_\nu^{1/q}(A_u) = R_{\nu q}(u), \qquad u \geq 0,$$

This gives, with our choice of q,

$$R_\nu(A_u) = [2R_{\nu/2}(u)]^{1/2}, \qquad u \geq 0. \tag{B.2}$$

Now we can write Z_1 as

$$Z_1 = \int_0^\infty dA_u \exp\left(-\frac{b^2 u}{2}\right) = \int_0^\infty \frac{du}{R_\nu^2(A_u)} \exp\left(-\frac{b^2 u}{2}\right). \tag{B.2'}$$

Using (5.1) we obtain

$$Z_1 = \frac{1}{2}\int_0^\infty \frac{du \exp(-b^2 u/2)}{R_{\nu/2}(u)}. \tag{B.3}$$

An Expression of the First Moment of Z. To compute the first moment of Z, we use Proposition 2.5:

$$E_z^{(m)}\left(\frac{1}{(R_t)^{2\theta}}\right) = \frac{1}{\Gamma(\theta)}\int_0^1 dh\, h^{\theta-1}(1-h)^{m-\theta}\left(\frac{1}{2t}\right)^\theta \exp\left(-\frac{zh}{2t}\right), \tag{B.4}$$

where we take $m = \nu/2, \theta = 1/2, z = (r_0/2)^2$. We then deduce from (B.3) and (B.4) that

$$E(Z) = \frac{\lambda}{2\Gamma(\theta)}\int_0^1 dh\, h^{\theta-1}(1-h)^{m-\theta}\int_0^\infty \frac{du}{(2u)^\theta}\exp\left(\frac{-zh}{2u}\right)\exp\left(\frac{-b^2 u}{2}\right) \tag{B.5}$$

$$= \frac{\lambda}{2\sqrt{\pi}}\int_0^1 \frac{dh}{\sqrt{h}}(1-h)^{(\nu-1)/2}\int_0^\infty \frac{du}{\sqrt{2u}}\exp\left(\frac{-zh}{2u}\right)\exp\left(\frac{-b^2 u}{2}\right).$$

We could also have avoided the time change operation with the help of the following formula, which is an important example of a Laplace transform for quadratic functionals of Brownian motion (see Revuz and Yor 1991, Corollary 1.8, p. 414):

$$E_{r_0}^{(\nu)}\left[\exp -\frac{b^2}{2}\int_0^t ds X_s\right] = \exp\left(-\frac{r_0 b}{2}\text{th}(bt)\right)[\text{ch}(bt)]^{-(\nu+1)}. \tag{B.6}$$

(Again, we use the notation ch and th for hyperbolic cosine and tangent.)

This gives

$$E[Z_1] = \int_0^\infty \frac{dt}{(\operatorname{ch}(bt))^{\nu+1}} \exp\left(-\frac{r_0 b}{2} \operatorname{th}(bt)\right)$$
$$= \frac{1}{b} \int_0^\infty \frac{ds}{(\operatorname{ch}(s))^{\nu+1}} \exp\left(-\frac{r_0 b}{2} \operatorname{th}(s)\right).$$

Making the change of variables $h = \operatorname{th}^2 s$, we obtain

$$E[Z] = \frac{\lambda}{2b} \int_0^1 \frac{dh}{\sqrt{h}} (1-h)^{(\nu-1)/2} \exp(-b\sqrt{zh}). \qquad (B.7)$$

This is another expression for (B.5) as shown below.

We start using the classical integral representation for Bessel functions K_θ:

$$\left(\frac{b}{c}\right)^{\alpha-1} 2K_{\alpha-1}(bc) = \int_0^\infty \frac{dy}{y^\alpha} \exp -\frac{1}{2}\left(b^2 y + \frac{c^2}{y}\right),$$

(see, for instance, Lebedev 1972, p. 119), and the important particular cases $\alpha = 1/2$ and $\alpha = 3/2$:

$$K_{-1/2}(x) = K_{1/2}(x) = (\pi/2x)^{1/2} e^{-x}. \qquad (B.8)$$

Returning to (B.5) we obtain

$$E[Z_1] = \frac{1}{2\sqrt{\pi}} \int_0^1 \frac{dh}{\sqrt{h}} (1-h)^{(\nu-1)/2} \int_0^\infty \exp\left(\frac{-zh}{2u}\right) \exp\left(\frac{-b^2 u}{2}\right) \frac{du}{\sqrt{2u}};$$

therefore,

$$E[Z_1] = \frac{1}{2\sqrt{\pi}} \int_0^1 \frac{dh}{\sqrt{h}} (1-h)^{(\nu-1)/2} \frac{1}{\sqrt{2}} \left(\frac{b}{\sqrt{zh}}\right)^{-1/2} 2K_{1/2}(b\sqrt{zh})$$
$$= \frac{1}{2\sqrt{\pi}} \int_0^1 \frac{dh}{\sqrt{h}} (1-h)^{(\nu-1)/2} \frac{\sqrt{\pi}}{b} \exp(-b\sqrt{zh}),$$

using (B.7). Remembering that $E(Z) = \lambda E(Z_1)$, the last expression is precisely the righthand side of (B.7).

References

Bick, A. (1991). Quadratic Variation Based Dynamic Strategies. European Finance Assoc. Meeting, Rotterdam

Black, F. and Scholes, M. (1973). The Pricing of Options and Corporate Liabilities. *J. Political Econom.*, **81**, 637–654

Bouaziz, L., Bryis, E. and Crouhy, M. (1994). The Pricing of Forward-Starting Asian Options. *J. Banking and Finance*, **18**, 623–639

Bougerol, P. (1983). Exemples de Théorèmes Locaux sur les Groupes Résolubles. *Ann. Inst. H. Poincaré*, **19**, 369–391

Carverhill, A.P. and Clewlow, L.J. (1990). Valuing Average Rate (Asian) Options. *Risk*, **3**, 25–29

Cox, J., Ingersoll, J. and Ross, S. (1985). A Theory of the Term Structure of Interest Rates. *Econometrica*, **53**, 363–384

Dufresne, D. (1989). Weak Convergence of Random Growth Processes with Applications to Insurance. *Insurance: Math. Econom.*, **8**, 187–201

Dufresne, D. (1990). The Distribution of a Perpetuity, with Applications to Risk Theory and Pension Funding. *Scand. Actuarial J.*, 39–79

Feller, W. (1964). *An Introduction to Probability Theory and Its Applications.* **2**, New York: Wiley

Garman, M.B. and Kohlhagen, S.W. (1983). Foreign Currency Option Values. *J. Int. Money Finance*, **2**, 231–237

Geman, H. (1989). The Importance of the Forward-Neutral Probability Measure in a Stochastic Approach to Interest Rates E.S.S.E.C. Working paper

Geman, H. (1992). Asset Allocation, Portfolio Insurance and Synthetic Securities. *Appl. Stoch. Models Data Anal.*, **8**, 179–188

Geman, H. and Portait, R. (1989). A Framework for Interest Rate Risk Analysis and Portfolio Management. American Stock Exchange Colloq., New York

Geman, H. and Yor, M. (1992). Quelques relations entre processus de Bessel, options asiatiques, et fonctions confluentes hypergéométriques. *C. R. Acad. Sci. Paris Sér. I*, 471–474. **Paper [3] in this book**

Harrison, J.M. and Kreps, D. (1979). Martingale and Arbitrage in Multiperiods Securities Markets. *J. Econom. Theory*, **20**, 381–408

Harrison, J.M. and Pliska, S.R. (1981). Martingales and Stochastic Integrals in the Theory of Continuous Trading. *Stoch. Proc. Appl.*, **11**, 215–260

Hill, J., Jain, A. and Wood, R.A. (1988). Portfolio Insurance: Volatility Risk and Futures Mispricing. *J. Portfolio Management*, Winter

Hull, J. and White, A. (1987). The Pricing of Options on Assets with Stochastic Volatilities. *J. Finance*, **42**, 281–300

Itô, K. and McKean, H.P. (1965). *Diffusion Processes and Their Sample Paths.* Berlin: Springer-Verlag

Kemna, A.G.Z. and Vorst, A.C.F. (1990). A Pricing Method for Options Based on Average Asset Values. *J. Banking Finance*, **14**, 113–129

Lebedev, N. (1972). *Special Functions and Their Applications.* New York: Dover

Lévy, E. (1992). Pricing European Average Rate Currency Options. *J. Int. Money Finance*, **11**, 474–491

Lévy, E. and Turnbull, S. (1992). Average Intelligence. *Risk*, **5**, 53–59

Longstaff, F. (1990). Pricing Options with Extendible Maturities: Analysis and Applications. *J. Finance*, **45**, 935–956

Pitman, J.W. and Yor, M. (1982). A Decomposition of Bessel Bridges. *Z. Wahrscheinlichkeitstheorie*, **59**, 425–457

Rendleman, R.J. and O'Brien, T.J. (1990). The Effects of Volatility Misestimation on Option Replication Portfolio Insurance. *Financial Anal. J.*, May–June, 60–61

Revuz, D. and Yor, M. (1991). *Continuous Martingales and Brownian Motion*. Berlin: Springer-Verlag

Rogers, L.C.G. and Williams, D. (1987). *Diffusions, Markov Processes and Martingales*. New York: Wiley

Ruttiens, A. (1990). Average-Rate Options, Classical Replica. *Risk*, **3**, 33–36

Shiga, T. and Watanabe, S. (1973). Bessel Diffusions as a One-Parameter Family of Diffusion Processes. *Z. Wahrscheinlichkeitstheorie*, **27**, 37–46

Turnbull, S. and Wakeman, L.M. (1991). A Quick Algorithm for Pricing European Average Options. *Financial Quant. Anal.*, **26**, 377–389

Vasicek, O. (1977). An Equilibrium Characterization of the Term Structure. *J. Financial Econom.*, **5**, 177–188

Vorst, T. (1992). Prices and hedge ratios of average exchange rate options. *Intern. Review of Financial Analysis*, **1** (3), 179–193

Williams, D. (1974). Path Decomposition and Continuity of Local Time for One-Dimensional Diffusions. *Proc. London Math. Soc.*, **28** (3), 738–768

Yor, M. (1980). Loi de l'indice du lacet brownien, et distribution de Hartman-Watson. *Z. Wahrscheinlichkeitstheorie*, **53**, 71–95

Yor, M. (1992a). *Some Aspects of Brownian Motion*. Part I: *Some Special Functionals*. Lecture Notes in Math., E.T.H. (Zürich). Basel: Birkhäuser

Yor, M. (1992b). On Some Exponential Functionals of Brownian Motion. *Adv. Appl. Probab.*, **24**, 509–531. **Paper [2] in this book**

Yor, M. (1992c). Sur les lois des fonctionnelles exponentielles du mouvement brownien, considérées en certains instants aléatoires. *C.R. Acad. Sci. Paris Sér. I*, **314**, 951–956. **Paper [4] in this book**

Postscript #5

A further discussion of this paper and perspectives in Mathematical finance are found in H. Geman's contribution **[0]** at the beginning of this monograph.

Further Results on Exponential Functionals
of Brownian Motion

1. Introduction

Let $(B_t, t \geq 0)$ denote a real-valued Brownian motion starting from 0, and let ν be a real.

The so-called geometric Brownian motion with drift ν, which is defined as:

$$\exp(B_t + \nu t), \ t \geq 0,$$

is often taken as the main stochastic model in Mathematical finance. This process may be represented as:

$$\exp(B_t + \nu t) = R^{(\nu)}_{A^{(\nu)}_t}, \quad t \geq 0, \tag{1.a}$$

with:

$$A^{(\nu)}_t = \int_0^t ds \exp 2(B_s + \nu s), \tag{1.b}$$

and $(R^{(\nu)}_u, u \geq 0)$ denotes a Bessel process with index ν, starting from 1. Both for theoretical reasons, and practical purposes, the law of the process $(A^{(\nu)}_t, t \geq 0)$, taken at various (possibly random) times has been of some interest in recent years.

In a previous paper [16], the following result was obtained: let T_λ be an exponential variable, with parameter λ, which is independent of $(B_t, t \geq 0)$; then, we have:

$$P(\exp(B^{(\nu)}_{T_\lambda}) \in d\rho, \ A^{(\nu)}_{T_\lambda} \in du) = \frac{\lambda}{2\rho^{2+\mu-\nu}} \ p^\mu_u(1, \rho)d\rho \, du \tag{1.c}$$

where $\mu = \sqrt{2\lambda + \nu^2}$, and $p^\mu_u(a, \rho)d\rho$ is the semi-group, taken at time u, of the Bessel process $(R^{(\mu)}_u, u \geq 0)$ starting from a.

A number of results about Bessel processes are presented in [16], among which the following formula for $p^\mu_u(a, \rho)$:

$$p^\mu_u(a, \rho) = \left(\frac{\rho}{a}\right)^\mu \frac{\rho}{u} \exp\left(-\frac{a^2 + \rho^2}{2u}\right) I_\mu\left(\frac{a\rho}{u}\right), \tag{1.d}$$

where I_μ denotes the modified Bessel function of index μ.

Using formula (1.d), together with the (implicit) Laplace transform in λ presented in (1.c), it is possible to obtain an expression for the joint law of $(\exp(B_t^{(\nu)}), A_t^{(\nu)})$, for a fixed time t. Indeed, taking up the notation in Section 6 of [16], we define $a_t(x, u)$ as:

$$P(A_t^{(\nu)} \in du \mid B_t^{(\nu)} = x) = a_t(x, u)du \qquad (1.e)$$

(as explained in [16], a does not depend on ν). Then, we have:

$$\frac{1}{\sqrt{2\pi t}} \exp\left(-\frac{x^2}{2t}\right) a_t(x, u) = \frac{1}{u} \exp\left(-\frac{1}{2u}(1 + e^{2x})\right) \theta_{e^x/u}(t) \qquad (1.f)$$

where $\theta_r(u)$ is characterized by the formula:

$$I_{|\nu|}(r) = \int_0^\infty \exp\left(-\frac{\nu^2 u}{2}\right) \theta_r(u)du \qquad (\nu \in \mathbb{R}, \ r > 0). \qquad (1.g)$$

An integral formula for $\theta_r(u)$ is presented in ([16], formula (6.b)), which in turn leads to an integral formula for the joint law of $(\exp(B_t^{(\nu)}), A_t^{(\nu)})$ (see [16], formula (6.e)). This result has just been used by Kawazu-Tanaka [8] to obtain some asymptotics for certain diffusions in random media. The same result may also be used to give an expression of the important quantity:

$$E[(A_t^{(\nu)} - k)^+] \qquad (1.h)$$

which governs the price of the so-called financial Asian options.

However, it appears that the expression thus obtained for (1.h) is too difficult for computational purposes and, at this point, it seems better to consider again the randomized functional: $A_{T_\lambda}^{(\nu)}$, which has a remarkably simple distribution, presented in Theorem 1 below. Consequently, the expression:

$$E[(A_{T_\lambda}^{(\nu)} - k)^+] = \lambda \int_0^\infty dt \, e^{-\lambda t} E[(A_t^{(\nu)} - k)^+]$$

has also a simple form (see Corollary 1.3 below) which, doubtless, it should be possible to use for numerical computations.

Theorem 1. Let $\lambda > 0$; define $\mu = \sqrt{2\lambda + \nu^2}$.
Then, if T_λ denotes a random time, which is exponentially distributed, with parameter λ, and is independent of B, one has:

$$A_{T_\lambda}^{(\nu)} \overset{(law)}{=} \frac{Z_{1,a}}{2Z_b}, \text{ where } a = \frac{\mu + \nu}{2}, \ b = \frac{\mu - \nu}{2} \qquad (1.i)$$

and $Z_{\alpha,\beta}$, resp. Z_γ, denotes a beta variable with parameters (α, β), resp. a gamma variable with parameter γ, that is:

$$P(Z_{\alpha,\beta} \in du) = \frac{u^{\alpha-1}(1-u)^{\beta-1}du}{B(\alpha, \beta)}, 0 < u < 1; \ P(Z_\gamma \in du) = \frac{e^{-u}u^{\gamma-1}du}{\Gamma(\gamma)},$$

and $Z_{1,a}$ and Z_b are further assumed to be independent.

We now illustrate Theorem 1 with the following Corollaries.

Corollary 1.1. *Let T be an exponentially distributed random variable, with parameter 1, i.e. $E(T) = 1$, which is assumed to be independent of the Brownian motion $(B_t, t \geq 0)$.*

Then, considering successively, in the statement of Theorem 1, the values $\nu = 0, \nu = 1/2$ and $\nu = -1/2$, one obtains:

$$\int_0^{T/2} ds \; \exp(B_s) \stackrel{(law)}{=} \frac{U}{4T}; \qquad \int_0^T ds \; \exp(2B_s + s) \stackrel{(law)}{=} U\sigma;$$

$$\int_0^T ds \; \exp(2B_s - s) \stackrel{(law)}{=} \frac{1 - U^2}{2T}, \tag{1.j}$$

where U is a uniform random variable valued in $[0, 1]$, $\sigma = \inf\{t : B_t = 1\}$, and the random variables which are featured on the right-hand sides in each of the three identities in law are assumed to be independent.

Corollary 1.2. *Let $\nu > 0$. Then, one has:*

$$\int_0^\infty ds \; \exp 2(B_s - \nu s) \stackrel{(law)}{=} \frac{1}{2Z_\nu}. \tag{1.k}$$

The identity in law (1.k) is easily deduced from (1.i), in which we let λ converge to 0; (1.k) was obtained first by D. Dufresne [3], and then reproved in [15], where the connection with last passage times for Bessel processes is established.

As announced before stating Theorem 1, we now present an expression of the Laplace transform (with respect to the variable t) of the quantity (1.h).

Corollary 1.3. *For every $\nu \geq 0$, $\lambda > 2(1 + \nu)$, and $k \geq 0$, one has:*

$$\lambda \int_0^\infty dt \; e^{-\lambda t} E[(A_t^{(\nu)} - k)^+] = \frac{\int_0^{1/2k} dt \; e^{-t} \, t^{\frac{\mu - \nu}{2} - 2}(1 - 2kt)^{\frac{\mu + \nu}{2} + 1}}{(\lambda - 2(1 + \nu)) \, \Gamma(\frac{\mu - \nu}{2} - 1)} \tag{1.l}$$

It has been pointed out to the author that the actual quantity of interest for financial Asian options is not so much (1.h), but:

$$E\left[\left(\frac{1}{t} A_t^{(\nu)} - k\right)^+\right] \equiv \frac{1}{t} E[(A_t^{(\nu)} - kt)^+]. \tag{1.h'}$$

Of course, if an explicit, simple expression for (1.h) were available, then, it would suffice to replace in such an expression the constant k by (kt), and then divide the obtained quantity by t. Since, despite (1.f) and (1.g), no such simple quantity has been obtained, we shall show, in Section 8, that, at the cost of a further Laplace transform (with respect to the variable k) an expression for (1.h') may be obtained.

1.3. We now give the details of the organization of the present paper:

- in Section 2, some prerequisites about Bessel processes, which complete those presented in [16], are given;
- in Section 3, two proofs of Theorem 1 are given, together with a partial explanation of the identity in law (1.i); furthermore, the arguments of the proofs of Theorem 1 lead to some other identities in law, presented at the end of Section 3;
- in Section 4, we obtain more identities in law, after first reproving Bougerol's identity in law (see [1]):

$$\text{for fixed } t \geq 0, \quad \sinh(B_t) \overset{\text{(law)}}{=} \gamma_{A_t}, \tag{1.m}$$

where $(\gamma_u, u \geq 0)$ is a Brownian motion starting from 0, which is assumed to be independent of B, and $A_t \equiv A_t^{(0)}$.

We already gave a (computational) proof of (1.m) in [16], using some classical integral formulae for Bessel functions; here, the proof of (1.m) relies upon our previous identity in law (1.i).

Then, with the help of (1.m), we are able to obtain the laws of variables of the form: A_S, where S varies amongst a family of random variables which are independent of B, and have some particular distributions;

- in Section 6, we use the conformal invariance of planar Brownian motion, and, more generally, the skew–product representation of Brownian motion in $\mathbb{R}^n$ $(n \geq 2)$ in order to obtain some more identities in law for variables of the form $A_S^{(\nu)}$, as we just described.
- in Section 7, we present some generalizations of the results obtained in the previous sections; indeed, both the Brownian motion with drift $(B_t + \nu t, t \geq 0)$ and the exponential function may be replaced by, respectively, some adequate diffusion, resp: function; again, such generalizations may have some important applications in Mathematical finance, when the archetype model of geometric Brownian motion is replaced by some other models;
- in Section 8, finally, we look at some computational issues which arise in the study of Asian options.

1.4. Some of the results discussed in this paper have been presented, without proof, in [17], whereas a detailed discussion of the implications in Mathematical finance is being made in Geman–Yor [5].

2. Some Complements on Bessel Processes

2.1. For the clarity of the exposition below, we need to take up the main part of the discussion on Bessel processes presented in Section 2. of [16], and to add some complements which will be used below.

Let Q_x^δ denote the law, on $C(\mathbb{R}_+, \mathbb{R}_+)$, of the square, starting from x, of a Bessel process with dimension δ; one of the main properties of the family $(Q_x^\delta; \delta > 0, x \geq 0)$ is the additivity property:

$$Q_{x+x'}^{\delta+\delta'} = Q_x^\delta * Q_{x'}^{\delta'} \qquad (\delta, \ \delta', x, x' \geq 0). \tag{2.a}$$

From this additivity property, we deduce the following important consequence: fix $t \geq 0$, and $u \geq 0$; then, the function of (x, δ):

$$q(x, \delta) \stackrel{\text{def}}{=} Q_x^\delta(\exp -t X_u)$$

is multiplicative in both arguments, i.e.:

$$q(x + x', \delta + \delta') = q(x, \delta) \, q(x', \delta') \qquad (x, x' \geq 0; \ \delta, \delta' \geq 0).$$

It is then easy to obtain:

$$Q_x^\delta(\exp(-t X_u)) = (1 + 2tu)^{-\delta/2} \exp\left(-x \frac{t}{1 + 2tu}\right) \tag{2.b}$$

by computing first this quantity for $\delta = 1$, say.

2.2. The following remarks about the absolute continuity relationship:

$$P_a^\nu \big|_{\mathcal{R}_t \cap (t < T_0)} = \left(\frac{R_t}{a}\right)^\nu \exp\left(-\frac{\nu^2}{2} \int_0^t \frac{ds}{R_s^2}\right) \cdot P_a^0 \big|_{\mathcal{R}_t} \tag{2.c}$$

which is valid for $a > 0$, and $\nu \in \mathbb{R}$, will also play an important role in the sequel. Here, in agreement with the notation in [16], Section 2, P_a^ν denotes the law on $C(\mathbb{R}_+, \mathbb{R}_+)$ of the Bessel process with index ν, starting from a, $(R_t)_{t \geq 0}$ is the process of coordinates, $\mathcal{R}_t = \sigma(R_s, s \leq t)$, and $T_0 = \inf\{t : R_t = 0\}$ $(\leq \infty)$.

As a consequence of (2.c), we remark that, for $\nu > 0$, we have:

$$P_a^{-\nu} \big|_{\mathcal{R}_t \cap (t < T_0)} = \left(\frac{a}{R_t}\right)^{2\nu} \cdot P_a^\nu \big|_{\mathcal{R}_t}, \tag{2.d}$$

and, as an application, we derive the identity:

$$P_a^{-\nu}(T_0 > t) = E_a^\nu\left(\left(\frac{a}{R_t}\right)^{2\nu}\right) \tag{2.e}$$

which demonstrates, if need be, that the $(P_a^\nu, (\mathcal{R}_t)_{t \geq 0})$ local martingale: $((\frac{1}{R_t})^{2\nu}, t \geq 0)$ is not a martingale.

2.3. It will also be interesting to consider the following time reversal result:

$$Under\ P_a^{-\nu},\ the\ process(R_{T_0-t};t\leq T_0)\ is\ distributed\ as$$
$$(R_t,t\leq L_a)\ under\ P_a^{\nu},\ where\ L_a=\sup\{t\geq 0:R_t=a\}. \tag{2.f}$$

Putting (2.e) and (2.f) together, we obtain the following

Proposition 1. *The common distribution of T_0, under $P_a^{-\nu}$, and of L_a, under P_0^{ν}, is that of: $a^2/2Z_{\nu}$.*

Proof.   i) The fact that T_0, under P_a^{ν}, and L_a, under P_0^{ν}, have the same law, follows from (2.f).

ii) In order to show that this common distribution is that of $a^2/2Z_{\nu}$, we now use the identity (2.e) and formula (2.b).

Using the elementary formula:

$$\frac{1}{x^{\nu}}=\frac{1}{\Gamma(\nu)}\int_0^{\infty}ds\ \exp(-sx)\ s^{\nu-1},$$

we deduce from (2.e) that:

$$P_a^{-\nu}(T_0>t)=a^{2\nu}\frac{1}{\Gamma(\nu)}\int_0^{\infty}ds\ s^{\nu-1}\ E_a^{\nu}(\exp-sR_t^2).$$

Then, applying formula (2.b) in the equivalent form:

$$E_a^{\nu}(\exp-sR_t^2)=\frac{1}{(1+2st)^{1+\nu}}\ \exp\left(-\frac{a^2s}{1+2st}\right),$$

one obtains the result stated in the proposition.   □

Remark. Another proof of the statement in Proposition 1 is given in [15], together with references to previous papers by Getoor and Pitman–Yor.

3.   Two Proofs of Theorem 1, and a Partial Explanation

3.1. In order to obtain the law of $A_{T_\lambda}^{(\nu)}$, we start with the following chain of equalities, which is a consequence of the time change relation (1.a):

$$P(A_{T_\lambda}^{(\nu)}\geq u)=\lambda\int_0^{\infty}dt\ e^{-\lambda t}\ P(A_t^{(\nu)}\geq u)$$
$$=\lambda\int_0^{\infty}dt\ e^{-\lambda t}P(H_u^{(\nu)}\leq t)=E[\exp(-\lambda\ H_u^{(\nu)})],$$

where $H_u^{(\nu)}\stackrel{\text{def}}{=}\int_0^u\frac{ds}{(R_s^{(\nu)})^2}.$

Using the absolute continuity relationship (2.c), we obtain:

$$E[\exp(-\lambda H_u^{(\nu)})] = E_1^0 \left[(R_u)^\nu \exp\left(-\frac{\mu^2}{2} H_u\right)\right] = E_1^\mu \left[\frac{1}{(R_u)^{\mu-\nu}}\right],$$

so that the following equality holds:

$$P(A_{T_\lambda}^{(\nu)} \geq u) = E_1^\mu \left[\frac{1}{(R_u)^{\mu-\nu}}\right]. \tag{3.a}$$

We now obtain an integral representation of the right-hand side of formula (3.a).

Proposition 2. *Let $\mu \geq b \geq 0$. Then, for every $r \geq 0$, one has:*

$$E_r^\mu \left(\frac{1}{(R_u)^{2b}}\right) = \frac{1}{\Gamma(b)} \int_0^{1/2u} dv \; \exp(-r^2 v) v^{b-1}(1-2uv)^{\mu-b}. \tag{3.b}$$

Proof. It uses the same arguments as those found in the proof of Proposition 1. Precisely, from the elementary formula:

$$\frac{1}{x^b} = \frac{1}{\Gamma(b)} \int_0^\infty dt \; \exp(-tx) t^{b-1},$$

one deduces:

$$E_r^\mu \left(\frac{1}{(R_u)^{2b}}\right) = \frac{1}{\Gamma(b)} \int_0^\infty dt \; t^{b-1} E_r^\mu \left(\exp -tR_u^2\right).$$

We now write formula (2.b) in the equivalent form:

$$E_r^\mu \left[\exp(-tR_u^2)\right] = \frac{1}{(1+2tu)^{\mu+1}} \exp\left(-r^2 \frac{t}{1+2tu}\right),$$

and we finally obtain formula (3.b) by changing the variable t into: $v = \frac{t}{1+2tu}$.
□

Using formula (3.b), we remark that formula (3.a) may be written in the following form: using the notation $\hat{Z}_b$ for $1/2Z_b$, one gets:

$$P(A_{T_\lambda}^{(\nu)} \geq u) = E\left[\hat{Z}_b \geq u; \left(1 - \frac{u}{\hat{Z}_b}\right)^a\right]. \tag{3.c}$$

Now, if we use the notation: X for $Z_{1,a}$, we have:

$$P(X \geq x) = (1-x)^a$$

and we obtain, assuming that X and $\hat{Z}_b$ are independent:

$$P(X\hat{Z}_b \geq u) = P\left(\hat{Z}_b \geq u; \ X \geq \frac{u}{\hat{Z}_b}\right) = E\left[\hat{Z}_b \geq u; \left(1 - \frac{u}{\hat{Z}_b}\right)^a\right].$$

Hence, we deduce from (3.c) that:

$$P(A_{T_\lambda}^{(\nu)} \geq u) = P(X\hat{Z}_b \geq u), \tag{3.d}$$

which proves the identity in law (1.i).

3.2. It is, in fact, possible to give another proof of the identity in law (1.i), without using the explicit Laplace transform formula (2.b).

First, we remark that, from formula (2.d), and formula (3.a), we have:

$$P(A_{T_\lambda}^{(\nu)} \geq u) = E_1^{-\mu}\left(T_0 \geq u; \frac{1}{(R_u)^{2(b-\mu)}}\right). \tag{3.e}$$

On the other hand, thanks to Proposition 1, the right-hand side of formula (3.c) is equal to:

$$E_1^{-\mu}\left((T_0 \geq u); (2(T_0 - u))^{\mu - b}\frac{\Gamma(\mu)}{\Gamma(b)}\right). \tag{3.f}$$

Hence, since formula (3.c) is equivalent to the identity in law (1.i), we must be able, in order to give a second proof of this identity in law, to pass directly from (3.e) to (3.f), that is: to prove the following identity:

$$E_1^{-\mu}((T_0 \geq u); (R_u)^{2(\mu - b)}) = E_1^{-\mu}\left((T_0 \geq u) \ (2(T_0 - u))^{\mu - b}\frac{\Gamma(\mu)}{\Gamma(b)}\right). \tag{3.g}$$

Indeed, the right-hand side of (3.g) is, thanks to the strong Markov property, equal to:

$$E_1^{-\mu}\left((T_0 \geq u); \ E_{R_u}^{-\mu}((2T_0)^{\mu - b})\frac{\Gamma(\mu)}{\Gamma(b)}\right).$$

and this latter expression is equal to the left-hand side of (3.g) since, from Proposition 1, we deduce:

$$E_r^{-\mu}\left((2T_0)^{\mu - b}\right)\frac{\Gamma(\mu)}{\Gamma(b)} = r^{2(\mu - b)}.$$

3.3. In order to illustrate the method of time-change and change of probability which we used in Subsection 3.1, we now prove two other identities in law

Proposition 3. *We keep the notation: $a = \frac{\mu+\nu}{2}$ and $b = \frac{\mu-\nu}{2}$. Then, we have:*

1) *for* $r \geq 1$, $P(\sup_{u \leq T_\lambda} \exp(B_u + \nu u) \geq r) = \dfrac{1}{r^{2b}}$, *i.e.*

$$\sup_{u \leq T_\lambda} \exp(B_u + \nu u) \overset{(law)}{=} \frac{1}{Z_{2b,1}}$$

2) *for* $\rho \leq 1$, $P(\inf_{u \leq T_\lambda} \exp(B_u + \nu u) \leq \rho) = \rho^{2a}$, *i.e.*

$$\inf_{u \leq T_\lambda} \exp(B_u + \nu u) \overset{(law)}{=} Z_{2a,1}.$$

Proof.   1) From (1.a), we have:

$$\sup_{u \leq T_\lambda} (\exp(B_u + \nu u)) = \sup_{s \leq A_{T_\lambda}^{(\nu)}} (R_s^{(\nu)}),$$

hence:

$$P\left\{ \sup_{u \leq T_\lambda} \exp(B_u + \nu u) \geq r \right\} = \lambda \int_0^\infty dt\, e^{-\lambda t}\, P\left\{ \sup_{s \leq A_t^{(\nu)}} (R_s^{(\nu)}) \geq r \right\}$$

$$= \lambda \int_0^\infty dt\, e^{-\lambda t} P\{A_t^{(\nu)} \geq T_r^{(\nu)}\}, \quad \text{where } T_r^{(\nu)} = \inf\{u : R_u^{(\nu)} = r\}$$

$$= \lambda \int_0^\infty dt\, e^{-\lambda t} P\{t \geq H_{T_r^{(\nu)}}^{(\nu)}\} = E[\exp(-\lambda\, H_{T_r^{(\nu)}}^{(\nu)})] = \frac{1}{r^{\mu-\nu}} = \frac{1}{r^{2b}},$$

from the absolute continuity relation (2.c).

2) Similarly, we have:

$$P\left\{ \inf_{u \leq T_\lambda} \exp(B_u + \nu u) \leq \rho \right\} = \lambda \int_0^\infty dt\, e^{-\lambda t} P\left\{ \inf_{s \leq A_t^{(\nu)}} (R_s^{(\nu)}) \geq \rho \right\}$$

$$= \lambda \int_0^\infty dt\, e^{-\lambda t} P(A_t^{(\nu)} \geq T_\rho^{(\nu)}) = \lambda \int_0^\infty dt\, e^{-\lambda t}\, P(t \leq H_{T_\rho}^{(\nu)}; T_\rho < \infty)$$

$$= E[\exp(-\lambda H_{T_\rho}^{(\nu)}); T_\rho < \infty] = \frac{1}{\rho^{2b}} P_1^\mu(T_\rho < \infty),$$

from the absolute continuity relationship (2.c).

Now, it is well-known that, since $(\dfrac{1}{(R_t)^{2\mu}}, t \geq 0)$ is a local martingale under P_1^μ, which converges to 0, as $t \longrightarrow \infty$, then:

$$\sup_{t \geq 0} \frac{1}{(R_t)^{2\mu}} \overset{(law)}{=} \frac{1}{U},$$

with U a uniformly distributed random variable on the interval $[0, 1]$; therefore, one has:

$$P_1^\mu (T_\rho < \infty) = P_1^\mu \left(\sup_{t \geq 0} \frac{1}{R_t^{2\mu}} \geq \frac{1}{\rho^{2\mu}} \right) = \rho^{2\mu},$$

and finally:

$$P \left\{ \inf_{u \leq T_\lambda} \exp(B_u + \nu u) \leq \rho \right\} = \frac{1}{\rho^{2b}} \rho^{2\mu} = \rho^{2a}. \qquad \square$$

3.4. In order to obtain a better understanding of the "factorization identity" (1.i), we now relate the law of $A_{T_\lambda}^{(\nu)}$ to that of the future supremum of a certain Bessel process with negative index.

Proposition 4. Let $\mu = (2\lambda + \nu^2)^{1/2}$, and $a = \frac{\mu + \nu}{2}$, $b = \frac{\mu - \nu}{2}$. Then, if we denote: $M = \sup_{v \leq T_0} R_v$, we have:

1) $P_r^{-\mu}(M \in dx) = (2\mu) \frac{r^{2\mu} \, dx}{x^{2\mu+1}} 1_{(x \geq r)},$

and, consequently:

$$E_r^{-\mu}[M^{2a}] = \frac{\mu}{b} r^{2a};$$

2) if we denote: $M_u = \sup_{u \leq v \leq T_0} R_v$, for $u \geq 0$, then, we have:

$$P(A_{T_\lambda}^{(\nu)} \geq u) = \frac{E_1^{-\mu}[(M_u)^{2a}]}{E_1^{-\mu}[M^{2a}]} \qquad (3.h)$$

3) if, for $x > 0$, we define: $L_x = \sup\{u : R_u = x\}$, then, we have, for $u \geq 0$:

$$P(A_{T_\lambda}^{(\nu)} \geq u) = \frac{E_1^{-\mu}[M^{2a}; \ L_{ZM} \geq u]}{E_1^{-\mu}[M^{2a}]}, \qquad (3.i)$$

where Z is independent of $(R_u \geq 0)$, and is a beta $(2a, 1)$ random variable,

i.e. $P(Z \in dx) = (2a)x^{2a-1}dx \qquad (0 < x < 1).$

3.5. Although the proof given in **3.2** is certainly much more illuminating than the proof given in **3.1**, we have not succeeded to obtain a more satisfactory explanation of the "factorization identity" (1.i).

Ideally, one would like to find out, for each λ and ν, two independent random variables N and D, which are measurable with respect to $\sigma\{(B_t, t \geq 0); T_\lambda\}$, and such that:

$$A_{T_\lambda}^{(\nu)} = \frac{N}{D}, N \overset{(\text{law})}{=} Z_{1,a}, \ D \overset{(\text{law})}{=} 2Z_b. \qquad (3.j)$$

However, it is not clear at all that this program may be fulfilled.

4. Some Applications of Theorem 1

4.1. Theorem 1 allows to obtain a quick proof of Bougerol's identity in law (1.m), which may be more natural than the proof given in [15].

Theorem 2. *(Bougerol [1]). Let $(B_t, t \geq 0)$ be a real-valued Brownian motion, starting from 0, and define $A_t = \int_0^t ds \, \exp(2B_s)$, $t \geq 0$. Then, we have:*

$$\text{for every fixed } t \geq 0, \quad \sinh(B_t) \overset{(law)}{=} \gamma_{A_t}. \tag{4.a}$$

Proof.   Let $\theta > 0$, and $\lambda = \frac{\theta^2}{2}$. In order to prove (4.a), it suffices, from the injectivity of the Laplace transform, to prove the equality:

$$E\left[|\sinh(B_{T_\lambda})|^\alpha\right] = E[|\gamma_{A_{T_\lambda}}|^\alpha], \tag{4.b}$$

for all sufficiently small α's.

It is well-known that $|B_{T_\lambda}|$ is an exponential random variable, with parameter θ. Hence, the left-hand side of (4.b) is equal to:

$$\theta \int_0^\infty dx \, \exp(-\theta x)(\sinh x)^\alpha, \tag{4.c}$$

whereas the right-hand side of (4.b) is equal to:

$$E\left(|N|^\alpha\right) E((A_{T_\lambda})^{\alpha/2}) \tag{4.d}$$

where N denotes a gaussian random variable, which is centered, and has variance 1.

Using jointly the duplication formula for the gamma function and the identity in law (1.i) for $\nu = 0$, it is easily shown that both quantities (4.c) and (4.d) have the common value:

$$\frac{1}{2^\alpha} B\left(\frac{\theta - \alpha}{2}, \alpha + 1\right),$$

which proves (4.b).   □

We have not been able to find an adequate extension of the identity in law (4.a) for $\nu \neq 0$, which would relate, say, the law of $B_t^{(\nu)}$ to that of $A_t^{(\nu)}$, for fixed t.

However, we have the following weaker relation

Corollary 2.1. *Let $\nu > 0$. Using the notation introduced in Theorem 1 and Theorem 2, we have the following identity in law:*

$$\gamma_{A_{T_\lambda}^{(\nu)}} \stackrel{(law)}{=} \frac{1}{\sqrt{Z_{b,\nu}}} \sinh(B_{T_{\lambda+\frac{\nu^2}{2}}}), \qquad (4.e)$$

where, on the right-hand side, $Z_{b,\nu}$ denotes a beta variable, with parameters (b,ν), which is assumed to be independent of the pair $\{(B_t; t \geq 0); T_{\lambda+\frac{\nu^2}{2}}\}$

Proof.　We remark that, from the algebraic relation between beta and gamma variables, we have:

$$Z_b \stackrel{(law)}{=} Z_{b,\nu} \, Z_a, \qquad (4.f)$$

where, on the right-hand side of (4.f), the variables $Z_{b,\nu}$ and Z_a are assumed to be independent.

Now, using the identity in law (1.i), we obtain:

$$A_{T_\lambda}^\nu \stackrel{(law)}{=} \frac{1}{Z_{b,\nu}} A_{T_{\lambda+\frac{\nu^2}{2}}},$$

so that, from the scaling property of Brownian motion, we deduce:

$$\gamma_{A_{T_\lambda}^{(\nu)}} \stackrel{(law)}{=} \frac{1}{\sqrt{Z_{b,\nu}}} \gamma_{A_{T_{\lambda+\frac{\nu^2}{2}}}} \stackrel{(law)}{=} \frac{1}{\sqrt{Z_{b,\nu}}} \sinh(B_{T_{\lambda+\frac{\nu^2}{2}}}), \text{ from } (4.a).$$

$$\square$$

4.2. We shall now exploit Bougerol's identity in law (4.a) in order to describe the laws of variables of the form A_T, when T varies amongst a fairly large class of random variables, assumed to be independent of $(B_t, t \geq 0)$. To do this, our main tool will be the following elementary

Lemma. *Let T be a strictly positive r.v., and g be the density of $\sinh(B_T) \stackrel{(law)}{=} \gamma_{A_T}$. Then, the law of B_T is given by:*

$$P(B_T \in dy) = dy(\cosh y) \, g(\sinh y).$$

Now, we write down a table, in which the main examples of distributions for T which we have found to be tractable are presented.

Table

$P(B_T \in dx)/dx$	$\frac{\theta}{2}\exp(-\theta\lvert x\rvert)$	$\dfrac{c_\alpha}{(\cosh (x))^\alpha}$	$\dfrac{a \cosh x}{\pi(a^2 + \sinh^2 x)}$	$\dfrac{x \coth x - 1}{(\sinh x)^2}$
A_T	$\dfrac{Z_{1,\theta/2}}{2Z_{\theta/2}}$	$\dfrac{1}{2Z_{\alpha/2}}$	σ_a	$\dfrac{1}{2U Z_1}$
T	$T_{\theta^2/2}$	$T^{(\alpha)}$	S_α	$T^{(3)}_{\pi/2} + \tilde{T}^{(3)}_{\pi/2}$

We now explain the Table, column after column:

- *first column*: this is simply a translation of formula (1.i), in the particular case $\nu = 0$, and $\lambda = \frac{\theta^2}{2}$;
- *second column*: here, α denotes any strictly positive real, and c_α is the normalizing constant which makes $\dfrac{c_\alpha}{(\cosh\ x)^\alpha}$ a density of probability on $\mathbb{R}$.

We find: $c_\alpha = \dfrac{\Gamma\left(\frac{\alpha+1}{2}\right)}{\sqrt{\pi}\,\Gamma\left(\frac{\alpha}{2}\right)}$; the random variable $T^{(\alpha)}$ satisfies:

$$E\left[\exp\left(-\frac{\lambda^2}{2}T^{(\alpha)}\right)\right] = c_\alpha \int_{-\infty}^{\infty} dx\ \exp(i\lambda x)\frac{1}{(\cosh\ x)^\alpha},$$

and we find:

$$E\left[\exp\left(-\frac{\lambda^2}{2}T^{(\alpha)}\right)\right] = \left|\frac{\Gamma\left(\frac{\alpha+i\lambda}{2}\right)}{\Gamma\left(\frac{\alpha}{2}\right)}\right|^2.$$

The cases $\alpha = 1$ and $\alpha = 2$ are particularly interesting; for $\alpha = 1$, $T^{(1)}$ may be represented as the first hitting time of $\pi/2$ by a reflecting Brownian motion starting from 0; for $\alpha = 2$, $T^{(2)}$ may be represented as the first hitting time of $\pi/2$ by a 3-dimensional Bessel process starting from 0.

- *third column*: this anticipates upon the discussion in Section 6, where the notation for S_α and σ_a are presented; the case $a = 1$ corresponds to $\alpha = 1$ in the second column.
- *fourth column*: $T^{(3)}_{\pi/2}$ and $\tilde{T}^{(3)}_{\pi/2}$ are two independent copies of the first hitting time of $\pi/2$ by BES(3), using the notation already introduced in the explanation of the second column.

5. Some Formulae Derived from Theorems 1 and 2

5.1. The following expression of the distribution of $A^{(\nu)}_{T_\lambda}$ is immediately deduced from Theorem 1:

$$P(A^{(\nu)}_{T_\lambda} \in dh) = \frac{dh}{2^b \Gamma(b)h^{b+1}}a\int_0^1 du\ (1-u)^{a-1}\ u^b\ e^{-u/2h}. \tag{5.a}$$

We remark that the density of the distribution of $A^{(\nu)}_{T_\lambda}$, featured in (5.a), may be expressed in terms of a confluent hypergeometric function:

5.2. During the Oberwolfach meeting on Mathematical finance, held in August 1992, D. Heath suggested that the results of Theorems 1 and 2 be used to obtain some closed form expression for the quantity:

$$E[\exp(-\alpha A^{(\nu)}_t)] \tag{5.c}$$

which plays an important role in the Black–Derman–Toy model ([19], [20], [21]). We start with the case $\nu = 0$, and use Bougerol's identity (4.a) in Theorem 2.

In this case, it is more convenient to take $\alpha = \frac{\xi^2}{2}$, for $\xi \in \mathbb{R}$, and it then follows from (4.a) that:

$$
E\left[\exp\left(-\frac{\xi^2}{2} A_t\right)\right] = \frac{1}{\sqrt{2\pi t}} \int_{-\infty}^{\infty} dx \, \exp\left(-\frac{x^2}{2t}\right) \exp(i\xi \, \sinh \, x)
$$

$$
= \sqrt{\frac{2}{\pi t}} \int_0^{\infty} dx \, \exp\left(-\frac{x^2}{2t}\right) \cos(\xi \, \sinh(x)). \qquad (5.d)
$$

In the general case: $\nu \in \mathbb{R}$, we are only able to deduce from Theorem 1 an expression of the Laplace transform in t of the quantity (5.c). Indeed, it follows from (5.a) that:

$$
E[\exp(-\alpha A_{T_\lambda}^{(\nu)})] = a \int_0^1 du \, (1-u)^{a-1} \, u^b \, I(u, \alpha), \qquad (5.e)
$$

where we have denoted:

$$
I(u, \alpha) = \int_0^{\infty} \frac{dh}{2^b h^{b+1} \Gamma(b)} \, \exp - \left(\frac{u}{2h} + \alpha h\right).
$$

We recall the classical integral representation for the modified Bessel function K_γ:

$$
K_\gamma(z) = \frac{1}{2}\left(\frac{z}{2}\right)^\gamma \int_0^{\infty} \frac{dt}{t^{\gamma+1}} \exp\left(-\left(t + \frac{z^2}{2t}\right)\right)
$$

(see, for example, Watson [13], p. 183, formula (14)), in terms of which we are able to express the quantity $I(u, \alpha)$:

$$
I(u, \alpha) = \frac{2}{\Gamma(b)} K_b(\sqrt{u\alpha}).
$$

Finally, we deduce from formula (5.e) the integral representation:

$$
E[\exp(-\alpha A_{T_\lambda}^{(\nu)})] = \frac{2a}{\Gamma(b)} \int_0^1 du \, (1 - u)^{a-1} \, u^b \, K_b(\sqrt{u\alpha}). \qquad (5.f)
$$

5.3. One of the purposes of Asian options is to take into account the fluctuation of the market over the time-interval $[0, t]$, hence the consideration of the L^1-mean:

$$
\frac{1}{t} A_t^{(\nu)} \equiv \frac{1}{t} \int_0^t ds \, \exp(2(B_s + \nu s)).
$$

In the same vein, one might consider instead the L^p-mean:

$$\left(\frac{1}{t}\int_0^t ds\ \exp(2p\left(B_s+\nu s\right))\right)^{1/p},\quad \text{for some } p \in [1,\infty[. \tag{5.g}$$

As $p \to \infty$, it follows from Laplace's asymptotic result that the expression in (5.g) converges almost surely towards

$$\sup_{s\le t}\ \exp(2(B_s+\nu s)). \tag{5.h}$$

Hence, the L^p-variant, resp: the L^∞-variant, of the computation (1.h) for the price of Asian options, consists in finding a closed formula, if possible, or, in estimating, the quantities:

$$E\left[\left(\int_0^t ds\ \exp\ 2p(B_s+\nu s)\right)^{1/p}\right] \tag{5.i$_p$}$$

and

$$E\left[\left\{\left(\sup_{s\le t}\ \exp(2(B_s+\nu s))\right)-k\right\}^{+}\right]. \tag{5.i$_\infty$}$$

Leaving this question aside for the moment, we check that Laplace's asymptotic result discussed above in our context is in agreement with the identity in law (1.i). For simplicity, we take $\nu = 0$, and $\lambda = \frac{\theta^2}{2}$.
As a particular case of the identity in law (1.i), we obtain:

$$\int_0^{T_\lambda} du\ \exp(2B_u)\ \stackrel{\text{(law)}}{=}\ \frac{1-U^{2/\theta}}{(2Z_{\theta/2})} \tag{5.j}$$

where T_λ is independent of B; on the right-hand side, U denotes a uniform random variable, and Z_a is a gamma variable with parameter a.

Using the scaling property of the law of BM, we deduce from (5.j) that:

$$\int_0^{T_\lambda} du\ \exp(2nB_u)$$

$$\stackrel{\text{(law)}}{=}\ \frac{1}{n^2}\int_0^{T_\lambda/n^2} du\ \exp(2B_u)\ \stackrel{\text{(law)}}{=}\ \frac{1}{n^2}\frac{(1-U^{2n/\theta})}{2Z_{(\theta/2n)}}. \tag{5.k}$$

Now, Laplace's asymptotic result tells us that, as $n \longrightarrow \infty$, we have:

$$\frac{1}{n}\log\left(\int_0^{T_\lambda} du\ \exp(2nB_u)\right)\ \xrightarrow[n\to\infty]{a.s.}\ 2\sup_{(u\le T_\lambda)}\ B_u \tag{5.l}$$

and the law of the right-hand side of (5.1) can be identified, since:

$$\sup_{u \leq T_\lambda} B_u \overset{\text{(law)}}{=} |B_{T_\lambda}|, \quad \text{and} \quad P(|B_{T_\lambda}| \in dx) = \theta \, e^{-\theta x} \, dx. \tag{5.m}$$

Hence, as a check on the identity in law (5.j), it may be interesting to prove the following convergence in law result for the right-hand side of (5.k):

$$\frac{1}{n} \log \left\{ \frac{1}{2n^2} \frac{1 - U^{2n/\theta}}{Z_{\theta/2n}} \right\} \overset{(d)}{\underset{n \to \infty}{\longrightarrow}} \left(\frac{2}{\theta} \right) T, \tag{5.n}$$

where T is a standard exponential variable (with parameter 1). [From (5.1) and (5.m), the right-hand side of (5.1) is distributed as $\frac{2}{\theta} T$].

Obviously, the convergence in law of the left-hand side of (5.n) amounts to that of:

$$\frac{1}{n} \log \left(\frac{1 - U^{2n/\theta}}{Z_{\theta/2n}} \right) = \frac{1}{n} \log(1 - U^{2n/\theta}) - \frac{1}{n} \log(Z_{\theta/2n}).$$

Since: $\log(1 - U^{2n/\theta}) \overset{a.s.}{\underset{n \to \infty}{\longrightarrow}} 0$, it remains to show, in order to prove (5.n) that:

$$-\varepsilon \, \log(Z_\varepsilon) \overset{(d)}{\underset{(\varepsilon \to 0)}{\longrightarrow}} T. \tag{5.o}$$

Proof of (5.o). Let $f : \mathbb{R} \longrightarrow \mathbb{R}_+$ be a bounded continuous function with compact support; we have:

$$\begin{aligned}
E\left[f\left(-\varepsilon \, \log \, Z_\varepsilon \right) \right] &= \frac{1}{\Gamma(\varepsilon)} \int_0^\infty dt \, e^{-t} \, t^{\varepsilon - 1} \, f(-\varepsilon(\log \, t)) \\
&= \frac{\varepsilon}{\Gamma(\varepsilon + 1)} \int_0^\infty dt \, t^{\varepsilon - 1} \, e^{-t} \, f(-(\log \, t^\varepsilon)) \\
&= \frac{1}{\Gamma(\varepsilon + 1)} \int_0^\infty du \, e^{(-u^{1/\varepsilon})} \, f(-\log \, u) \\
&\sim \int_0^\infty du \, e^{(-u^{1/\varepsilon})} \, f(-\log \, u) \sim \int_{-\infty}^\infty dt \, e^t \, e^{-(e^{t/\varepsilon})} \, f(-t)
\end{aligned}$$

The contribution of $\mathbb{R}_-$ to the last integral is equal to:

$$\int_0^\infty du \, e^{-u} \, e^{-(e^{-u/\varepsilon})} \, f(u) \underset{(\varepsilon \to 0)}{\longrightarrow} \int_0^\infty du \, e^{-u} \, f(u),$$

whereas it is equally easy to show that the contribution of $\mathbb{R}_+$ to the integral is negligible, as $\varepsilon \to 0$. Hence, the proof of (5.o) is complete.

5.4. The main unsatisfactory feature of Theorem 1, or equivalently of formula (5.a), is that it involves a Laplace transform in t.

We now try to invert this Laplace transform, or rather, we indicate some close relation with the Laplace transforms of some probability distributions on $\mathbb{R}_+$ which are associated to the reciprocal of the gamma function.

For simplicity, we take, in formula (5.a), $\nu = 0$; it will also be convenient to consider $\lambda = \frac{\theta^2}{2}$; in this case, we have: $a = b = \frac{\theta}{2}$, so that multiplying and dividing the right-hand side of (5.a) by θ, we obtain:

$$
\int_0^\infty dt \, \exp\left(-\frac{\theta^2 t}{2}\right) P\left(A_t \in dh\right)
$$

$$
= \frac{dh}{(2h)^{\frac{\theta}{2}+1} \, \Gamma\left(\frac{\theta}{2}+1\right)} \int_0^1 \frac{du}{1-u} e^{-u/2h} (u(1-u))^{\frac{\theta}{2}} \tag{5.a$'$}
$$

Bringing together the different quantities on the right-hand side of (5.a$'$) which depend on θ, one sees clearly that the Laplace transform in (5.a$'$) is closely related to the following

Proposition 5. *(P. Hartman [6]).*
Let $m > 0$; then, the function:

$$
H_m(\theta) = \frac{m^\theta}{\Gamma(1+\theta)} \qquad (\theta \ge 0)
$$

is the Laplace transform, in $\frac{\theta^2}{2}$, of a probability distribution μ_m on $\mathbb{R}_+$, i.e.

$$
\frac{m^\theta}{\Gamma(1+\theta)} = \int_0^\infty \exp\left(-\frac{\theta^2 t}{2}\right) \mu_m(dt) \tag{5.p}
$$

if, and only if: $m \le e^{-\gamma}$, where γ denotes Euler's constant.
When this condition on m is satisfied, there exists a density g_m such that:

$$
\mu_m(dt) = g_m(t)dt.
$$

For completeness, we shall give a proof of Proposition 5 below, but, for the moment, we show how, for h large enough, we are able to deduce from Proposition 5, an expression of $P(A_t \in dh)/dh$.

Indeed, from (5.p), we remark that, for $m \le e^{-\gamma}$:

$$
\frac{m^{\theta/2}}{\Gamma\left(1+\frac{\theta}{2}\right)} = \int_0^\infty \exp\left(-\frac{\theta^2 t}{2}\right) \tilde{g}_m(t)dt, \quad \text{where } \tilde{g}_m(t) = 4g_m(4t).
$$

Applying this result to $m = \frac{u(1-u)}{2h}$ in (5.a') (the condition on m, found in Proposition 5, is then satisfied if: $h \geq \frac{e^\gamma}{8}$), we obtain:

$$P\left(A_t \in dh\right) = \frac{dh}{2h} \int_0^1 \frac{du}{(1-u)} e^{-u/2h}\, \tilde{g}_{\frac{u(1-u)}{2h}}(t), \text{ for } h \geq \frac{e^\gamma}{8}. \qquad (5.q)$$

As a consequence of formula (5.q), we are able to express the important quantity:

$$E[(A_t - k)^+] \qquad\qquad \text{for } k \geq \frac{e^\gamma}{8}$$

in terms of the densities $(\tilde{g}_m(t), \ m \leq e^{-\gamma})$.

Indeed, we have, for $k \geq e^\gamma/8$:

$$E[(A_t - k)^+] = \int_k^\infty \frac{dh}{2h}(h-k) \int_0^1 \frac{du}{(1-u)} e^{-u/2h}\, \tilde{g}_{\frac{u(1-u)}{2h}}(t)$$

$$= \int_0^{1/k} \frac{dx}{2x}\left(\frac{1}{x}-k\right) \int_0^1 \frac{du}{(1-u)} e^{-\frac{ux}{2}}\, \tilde{g}_{\frac{u(1-u)x}{2}}(t)$$

$$= \int_0^1 \frac{du}{(1-u)} \int_0^{1/k} \frac{dx}{2x}\left(\frac{1}{x}-k\right) e^{-\frac{ux}{2}}\, \tilde{g}_{\left(\frac{u(1-u)x}{2}\right)}(t).$$

To be complete, we now give a

Proof of Proposition 5.

i) We first prove the result (5.p) for $m = e^{-\gamma}$.

In this case, we apply Weierstraß infinite product representation of the gamma function:

$$\frac{e^{-\gamma\theta}}{\Gamma(1+\theta)} = \prod_{j=1}^\infty \left(1+\frac{\theta}{j}\right) e^{-\frac{\theta}{j}} \qquad (*)$$

(see, for example, S.J. Patterson [11], p. 131).

Now, there exists a random variable X, valued in $\mathbb{R}_+$ such that:

$$(1+\theta)e^{-\theta} = E\left[\exp\left(-\frac{\theta^2}{2}X\right)\right]$$

and it is easily shown that:

$$P(X \in dx) = \frac{dx}{\sqrt{2\pi}}\left(\frac{e^{-\frac{1}{2x}}}{x^{5/2}}\right)$$

in other words, one has: $X \stackrel{(\text{law})}{=} \frac{1}{2Z_{3/2}}$, where we use the notation Z_γ for a gamma variable with parameter γ, i.e.:

$$P\left(Z_\gamma \in dz\right) = \frac{dz}{\Gamma(\gamma)} z^{\gamma-1} e^{-z} \qquad (z > 0).$$

Now, in order to represent the left-hand side of $(*)$ as a Laplace transform in $\frac{\theta^2}{2}$, we consider a sequence $(X_j; j = 1, 2, \dots)$ of i.i.d. random variables, with common distribution that of X.

With the help of this sequence, the right-hand side of $(*)$ may be represented as: $E[\exp - \frac{\theta^2}{2} \sum_{j=1}^\infty (\frac{1}{j^2} X_j)]$ so that the left hand side of $(*)$ now appears as the Laplace transform, in $\frac{\theta^2}{2}$, of the distribution of:

$$Y \stackrel{\text{def}}{=} \sum_{j=1}^\infty \frac{1}{j^2} X_j.$$

ii) In the case: $m < e^{-\gamma}$, we write:

$$\frac{m^\theta}{\Gamma(1+\theta)} = \left(\frac{m}{e^{-\gamma}}\right)^\theta \left(\frac{e^{-\gamma\theta}}{\Gamma(1+\theta)}\right), \qquad (**)$$

and we use the fact that for $a \equiv \log(\frac{e^{-\gamma}}{m}) > 0$,

$$\left(\frac{m}{e^{-\gamma}}\right)^\theta = \exp(-\theta a) = E\left[\exp\left(-\frac{\theta^2}{2} T_a\right)\right]$$

where T_a denotes here the first hitting time of a by 1-dimensional BM starting from 0; as is well-known, one has:

$$P\left(T_a \in dt\right) = \frac{a\, dt}{\sqrt{2\pi t^3}} \exp\left(-\frac{a^2}{2t}\right).$$

Hence, in this case, the left-hand side of $(**)$ appears as the Laplace transform of the distribution of: $T_a + Y$, T_a and Y being assumed to be independent.

iii) In the case: $m > e^{-\gamma}$, the function: $\theta \longrightarrow \frac{m^\theta}{\Gamma(1+\theta)}$ is increasing in a neighborhood of $\theta = 0_+$; hence, it cannot be a Laplace transform in $\frac{\theta^2}{2}$.

Remark. A probabilistic discussion of Weierstraß infinite product formula for the gamma function has recently been made by Fuchs and Letta [4]; see also Gordon [23].

6. Some Applications of the Conformal Invariance and Skew-product Representation of Brownian Motion in $\mathbb{R}^n$, $n \geq 2$

6.1. Let $Z_t = X_t + iY_t, t \geq 0$, be a complex valued Brownian motion, i.e.: $(X_t, t \geq 0)$ and $(Y_t, t \geq 0)$ are two independent real-valued Brownian motions.

P. Lévy [10] remarked that, as a consequence of the conformal invariance of the distribution of Z, if $f: \mathbb{C} \longrightarrow \mathbb{C}$ is an entire function, i.e. f is holomorphic on the whole complex plane $\mathbb{C}$, and f is not constant, then there exists another complex Brownian motion $(\hat{Z}(u), u \geq 0)$ such that:

$$f(Z_t) = \hat{Z}\left(\int_0^t ds \mid f'(Z_s) \mid^2 \right), \quad t \geq 0. \tag{6.a}$$

Several important consequences of this result of Lévy have now been obtained; see, for example, B. Davis [2], and in another direction, Pitman-Yor [12] who are concerned with the level crossings of a Cauchy process.

In the particular case where $f(z) = \exp(z)$, the equality (6.a) becomes:

$$\exp(Z_t) = \hat{Z}\left(\int_0^t ds \ \exp(2X_s) \right), \quad t \geq 0, \tag{6.b}$$

from which we deduce the following identities in law.

Theorem 3. Let $(B_t, t \geq 0)$ be a real valued Brownian motion, starting from 0; define, for $a > 0$, $\sigma_a = \inf\{t : B_t = a\}$.

Let $Z_t = X_t + iY_t, t \geq 0$, be a complex Brownian motion, starting from 0, and define $A_t = \int_0^t ds \ \exp(2X_s), t \geq 0$.

1) If $S = \inf\{t : |Y_t| = \frac{\pi}{2}\}$, then, we have:

$$A_S \overset{(law)}{=} \sigma_1; \tag{6.c}$$

2) More generally, to any $\alpha \in \mathbb{R}$, we associate: $a = (1 + \alpha^2)^{-1/2}$, and θ the unique real in $\left]-\frac{\pi}{2}, \frac{\pi}{2}\right]$ such that: $\tan(\theta) = \frac{1}{\alpha}$.
   Then, if $S_\alpha = \inf\{t : \cos(Y_t) = \alpha \ \sin(Y_t)\} \equiv \inf\{t : Y_t = \theta, \text{ or } \theta - \pi\}$, we have:

$$A_{S_\alpha} = \inf\{t : \hat{X}_t - \alpha\hat{Y}_t = 0\} \overset{(law)}{=} \sigma_a \equiv \inf\{t : B_t = a\}. \tag{6.d}$$

The proof follows immediately from the representation (6.b), and the elementary formula: $\exp(z) = \exp(x) \ (\cos \ y + i \ \sin \ y)$.

6.2. We now consider, more generally, $(Z_t, t \geq 0)$ a Brownian motion valued in $\mathbb{R}^n$, $n \geq 2$, and starting from $z_0 \neq 0$. For simplicity, we shall assume that $|z_0| = 1$.

As is now well-known (see Itô – Mc Kean [7], p. 270 and sq.), $(Z_t, t \geq 0)$ may be represented in the skew-product form:

$$Z_t = |Z_t| \, V(H_t), \ t \geq 0, \tag{6.e}$$

where $H_t = \int_0^t \frac{ds}{|Z_s|^2}$, $t \geq 0$, and $(V(u), u \geq 0)$ is a standard Brownian motion on S_{n-1} the unit sphere in $\mathbb{R}^n$, and V is independent of $(|Z_t|, t \geq 0)$. We may also represent the radial part of Z, i.e. $R_t = |Z_t|$, as:

$$R_t = \exp(B_u + \nu u)|_{u=H_t}, \ t \geq 0, \tag{6.f}$$

where $(B_u, u \geq 0)$ is a real-valued Brownian motion, starting from 0.
This latter representation (6.f) is nothing else but the representation (1.a) we started with, once we have remarked that:

$$H_t = \inf\left\{ u : A_u^{(\nu)} \equiv \int_0^u ds \ \exp \ 2(B_s + \nu s) > t \right\}. \tag{6.g}$$

We now replace in (6.e) the time variable t by $A_u^{(\nu)}$, which gives, thanks to (6.f):

$$\exp(B_u + \nu u) \, V(u) = Z_{A_u^{(\nu)}}, \ u \geq 0. \tag{6.h}$$

On the left-hand side of (6.h), the processes $(B_u; u \geq 0)$ and $(V(u); u \geq 0)$ are independent, whereas, on the right-hand side of (6.h), $A_u^{(\nu)}$ is measurable with respect to $(|Z_u|, u \geq 0)$, as it follows from (6.g). We now have the following extension of Theorem 3.

Theorem 4. *Let $\theta \in \mathbb{R}^n$, $|\theta| = 1$, and define:*
$$S_\theta = \inf\{u : (\theta, V(u)) = 0\},$$

where (h, k) denotes the euclidian scalar product between h and k.
Then, we have:

$$A_{S_\theta}^{(\nu)} \overset{(law)}{=} \sigma_{a_\theta},$$

where $a_\theta = (\theta, z_0)$, and σ_b has the same meaning as in Theorem 3.

The proof is just as immediate as that of Theorem 3, using the fact that $\{(\theta, Z_u), u \geq 0\}$ is a one-dimensional Brownian motion, starting from a_θ. A natural question is now to find out the distribution of S_θ.

7.   An Extension to Some Other Diffusions

The purpose of this paragraph is two-fold:

- on one hand, we are interested in some generalizations of Theorem 1 and its Corollaries when the Brownian motion with constant drift: $(B_t + \nu t; t \geq 0)$ is replaced by a diffusion which, for simplicity, is assumed to be a solution of the following equation:

$$X_t = 1 + B_t + \int_0^t ds \ h(X_s); \tag{7.a}$$

- on the other hand, we would like to understand better the method which led to Theorem 1, and one way to achieve this is to examine which kinds of difficulties we encounter when we cannot rely upon the special properties of (squares of) Bessel processes.

7.1. To begin with, we shall try to compute the law of

$$A_t = \int_0^t ds \ \exp(2X_s),$$

when (X_t) solves (7.a). In fact, just as in the case of the constant drift, i.e. $h(x) \equiv \nu$, it will be much easier to describe the law of A_{T_λ}, when T_λ is an independent exponential variable, with parameter λ.

We first apply Itô's formula to $(\exp(X_t), t \geq 0)$:

$$\exp(X_t) = 1 + \int_0^t \exp(X_s)dB_s + \int_0^t ds \left\{ \exp(X_s)h(X_s) + \frac{1}{2}\exp(X_s) \right\}. \tag{7.b}$$

We can write the last (Riemann) integral as:

$$\int_0^t dA_s \left(\frac{2h(X_s) + 1}{2\exp(X_s)} \right).$$

We now assume (for simplicity) that: $A_\infty = \infty$ a.s.; define $\tau_t = \inf\{u : A_u > t\}$, and the process $(R_t, t \geq 0)$ by the formula:

$$R_t = \exp(X_{\tau_t}), \ t \geq 0; \text{ then, we have: } \tau_t = \int_0^t \frac{ds}{R_s^2}.$$

Next, we deduce from (7.b) that $(R_t, t \geq 0)$ solves the following equation:

$$R_t = 1 + \beta_t + \int_0^t du \left(\frac{2h(\log R_u) + 1}{2R_u} \right), \tag{7.c}$$

and, in order to find out the law of A_{T_λ}, we write:

$$P\left(A_{T_\lambda} > u\right) = \lambda \int_0^\infty dt \ \exp(-\lambda t) P\left(A_t > u\right) = E\left[\exp(-\lambda \tau_u)\right] \qquad (7.d)$$

We now look for a function $\varphi_\lambda : \mathbb{R}_+ \longrightarrow \mathbb{R}_+$ such that: $\varphi_\lambda(1) = 1$, and:

$$\{\varphi_\lambda(R_u) \ \exp(-\lambda \tau_u), u \geq 0\} \quad \text{is a martingale}$$

with respect to the filtration $(\mathcal{R}_t \equiv \sigma\{R_u; u \leq t\}; \ t \geq 0)$.

Assume for a moment that we have found such a function φ_λ; we consider a new probability measure P^λ defined by:

$$P^\lambda_{|\mathcal{R}_u} = \varphi_\lambda(R_u) \ \exp(-\lambda \tau_u) \cdot P_{|\mathcal{R}_u}.$$

Going back to (7.d), we now remark that:

$$P\left(A_{T_\lambda} > u\right) = E^\lambda\left[\frac{1}{\varphi_\lambda(R_u)}\right]. \qquad (7.e)$$

We now translate the above discussion in terms of infinitesimal generators: if L denotes the infinitesimal generator of our original diffusion $(X_t, t \geq 0)$, then, we have:

$$Lf(x) = \frac{1}{2}f''(x) + h(x) \ f'(x), \qquad \text{for } f \in C_b^2(\mathbb{R}).$$

Now, $(R_t, t \geq 0)$ admits the infinitesimal generator $\tilde{L}$, which satisfies:

$$\tilde{L}g(r) = \frac{1}{2}g''(r) + \left(\frac{2h(\log \ r) + 1}{2r}\right) \ g'(r), \text{ for } g \in C_b^2(\,]0, \infty[\,)$$

and φ_λ satisfies the martingale condition demanded above if:

$$\tilde{L}\varphi_\lambda(r) = \frac{\lambda \varphi_\lambda(r)}{r^2}. \qquad (7.f)$$

We also remark that, under P^λ, $(R_t, t \geq 0)$ is a diffusion with generator:

$$\tilde{L}_\lambda = \tilde{L} + \left(\frac{\varphi_\lambda'}{\varphi_\lambda}\right)\frac{d}{dr}. \qquad (7.g)$$

With the help of formula (7.e), we now obtain the following expression of the density of A_{T_λ}:

$$P(A_{T_\lambda} \in du) = du \ E^\lambda\left[\tilde{L}_\lambda\left(\frac{1}{\varphi_\lambda}\right)(R_u)\right]. \qquad (7.h)$$

The same method leads to an expression for the joint law of $(\exp(X_{T_\lambda}); A_{T_\lambda})$;
Indeed, using our previous notation, we have:

$$E\left[f\left(\exp(X_{T_\lambda})\right); A_{T_\lambda} > u\right] = \lambda \int_0^\infty dt\, e^{-\lambda t}\, E\left[f\left(\exp(X_t)\right); A_t > u\right]$$

$$= \lambda\, E\left[\int_{\tau_u}^\infty dt\, e^{-\lambda t}\, f\left(\exp(X_t)\right)\right]$$

$$= \lambda\, E\left[\int_u^\infty d\tau_s\, \exp(-\lambda\tau_s)\, f(R_s)\right] = \lambda\, E^\lambda\left[\int_0^\infty \frac{ds}{R_s^2}\, \frac{1}{\varphi_\lambda(R_s)}\, f(R_s)\right].$$

Hence, if we denote by $P_t^\lambda(r, d\rho) \equiv p_t^\lambda(r, \rho)d\rho$ the semi group of the diffusion,
$(R_t, t \geq 0)$ under $(P_r^\lambda; r > 0)$, it follows from the above series of equalities
that:

$$E\left[f\left(\exp(X_{T_\lambda})\right); A_{T_\lambda} > u\right] = \lambda \int_0^\infty ds \int d\rho\, p_s^\lambda(1, \rho)\frac{1}{\rho^2 \varphi_\lambda(\rho)} f(\rho).$$

Consequently, we have obtained the following:

$$P\left(\exp(X_{T_\lambda}) \in d\rho; A_{T_\lambda} \in du\right) = \lambda\, p_u^\lambda(1, \rho)\frac{du\, d\rho}{\rho^2 \varphi_\lambda(\rho)}. \qquad (7.\text{i})$$

Therefore, introducing the notation:

$$v_\lambda(\rho) = \int_0^\infty du\, p_u^\lambda(1, \rho)$$

for the density of the resolvent of $(R_u, u \geq 0)$ under P^λ, we may write $(7.\text{i})$
as:

$$P\left(\exp(X_{T_\lambda}) \in d\rho;\ A_{T_\lambda} \in du\right) = \left(\frac{\lambda v_\lambda(\rho)d\rho}{\rho^2 \varphi_\lambda(\rho)}\right)\left(\frac{1}{v_\lambda(\rho)}p_u^\lambda(1, \rho)du\right), \qquad (7.\text{i}')$$

or, equivalently:

$$P(\exp(X_{T_\lambda}) \in d\rho) = \frac{\lambda v_\lambda(\rho)d\rho}{\rho^2 \varphi_\lambda(\rho)}, \qquad (7.\text{j})$$

and: $$P(A_{T_\lambda} \in du \mid \exp(X_{T_\lambda}) = \rho) = \frac{p_u^\lambda(1, \rho)du}{v_\lambda(\rho)}.$$

7.2. The method developed in paragraph (7.1) may be used to compute the distribution of A_{T_λ}, in case $(X_t, t \geq 0)$, the solution of (7.a), is such that the process $(R_t, t \geq 0)$, which solves (7.c) is the norm of a δ-dimensional Ornstein – Uhlenbeck process; more precisely, we consider $(Z_t, t \geq 0)$ an $\mathbb{R}_+$-valued diffusion with infinitesimal generator:

$$2z \frac{d^2}{dz^2} + (-2\theta z + \delta) \frac{d}{dz}, \qquad \text{for some } \theta, \delta > 0, \qquad \text{(7.k)}$$

and $R_t = (Z_t)^{1/2}$, $t \geq 0$, is assumed to satisfy $R_0 = Z_0 = 1$.

A number of useful results about the laws of the processes $(Z_t, t \geq 0)$ defined above are collected in Section 6 of Pitman-Yor [13]. Following the notation in [13], we denote the law of $(Z_t, t \geq 0)$ on $C(\mathbb{R}_+, \mathbb{R}_+)$ by $^{-\theta}Q_1^\delta$; we also use $(Z_t, t \geq 0)$ to denote the coordinate process on $C(\mathbb{R}_+, \mathbb{R}_+)$.

For the moment, we remark that, in this case, the corresponding process $(X_t, t \geq 0)$ which solves (7.a), admits the following drift function h:

$$h(x) = -\theta \, \exp(2x) + \nu,$$

where $\nu = \frac{\delta - 2}{2}$; in the particular case $\theta = 0$, $(X_t, t \geq 0)$ is a Brownian motion with constant drift ν.

In agreement with our notation throughout this paper, we should denote:

$$^{-\theta}A_t^{(\nu)} = \int_0^t ds \, \exp(2X_s).$$

Then, if we combine the method developed in Subsection 7.1 with the absolute continuity relation, which is a consequence of Girsanov's theorem, between $^{-\theta}Q_1^\delta$ and $Q_1^\delta = {}^0Q_1^\delta$, and if, moreover, we use the well-known fact that the process $(Z_u, u \geq 0)$ is distributed under $^{-\theta}Q_1^\delta$ as the process

$$(e^{-2\theta u} Z_{g_\theta(u)}, \, u \geq 0) \quad \text{under} \quad Q_1^\delta, \quad \text{where: } g_\theta(u) = \frac{e^{2\theta u} - 1}{2\theta},$$

it is not difficult to obtain the identity in law:

$$g_\theta(^{-\theta}A_{T_\lambda}^{(\nu)}) \overset{\text{(law)}}{=} A_{T_\lambda}^{(\nu)}$$

or, equivalently, using the injectivity of the Laplace transform:

$$\text{for } t > 0, \; g_\theta(^{-\theta}A_t^{(\nu)}) \overset{\text{(law)}}{=} A_t^{(\nu)}.$$

In fact, there is no need to have recourse to Subsection (7.1), and the last written identity in law may be considered as a trivial one, with the help of the following remarks:

consider $(X_t, t \geq 0)$, the solution of the equation:

$$X_t = B_t + \nu t - \theta \int_0^t ds \, \exp(2X_s), \qquad t \geq 0.$$

i.e. this is equation (7.a), where: $h(x) = \nu - \theta \exp(2x)$.

Then, the following *almost sure* identity holds:

$$g_\theta \left(\int_0^t ds \, \exp(2X_s) \right) = \int_0^t ds \, \exp 2(B_s + \nu s), \quad t \geq 0. \qquad (7.1)$$

Indeed, to prove (7.1), all we need to show is that the derivatives of both sides, with respect to t, are equal, i.e.:

$$\exp(2X_t) \exp\left(2\theta \int_0^t ds \, \exp(2X_s) \right) = \exp 2(B_t + \nu t), \quad t \geq 0.$$

But, taking the logarithms of both sides, we see that the last equality reduces precisely to the stochastic equation above, from which we defined the process $(X_t, t \geq 0)$.

7.3. To illustrate further the method developed in Subsection **7.1**, we now show how it may lead to the knowledge of the distribution of:

$$C_t = \int_0^t ds \, (\cosh(B_s))^2 \quad \text{and/or: } S_t = \int_0^t ds \, (\sinh(B_s))^2,$$

or, in fact, as above, to the distributions of these functionals taken at an independent exponential time T_λ, with parameter λ.

Before we develop some computations which parallel those made in Subsection **7.1**, we make some elementary remarks about the functionals $(C_t, t \geq 0)$ and $(S_t, t \geq 0)$:

– first, using the relation: $\cosh^2(x) - \sinh^2(x) = 1$, we obtain:

$$E[\exp(-\xi C_{T_\lambda})] = \frac{\lambda}{(\lambda + \xi)} \, E[\exp(-\xi S_{T_{\lambda+\xi}})];$$

hence, the knowledge of the distribution of C_{T_λ}, as λ varies in $\mathbb{R}_+$, entails that of S_{T_μ}, for $\mu > 0$;

– secondly, using the duplication formula: $\cosh(2x) = 2(\cosh(x))^2 - 1$, we obtain:

$$E\left[\exp(-\xi C_{T_\lambda})\right] = \lambda \int_0^\infty dt \, \exp\left(-\left(\lambda + \frac{\xi}{2}\right)t\right) \dots$$

$$\dots E\left[\exp -\frac{\xi}{2} \int_0^t ds \, (\cosh(2B_s))\right]$$

so that the knowledge of the distribution of C_{T_λ}, for $\lambda > 0$, entails that of:

$$\int_0^{T_\mu} ds \, (\cosh(2B_s)), \quad \text{for } \mu > 0,$$

and, using the scaling property of $(B_s, s \geq 0)$, we then have access to:

$$\int_0^{T_\mu} ds \, (\cosh(mB_s)), \quad \text{for any } m \in \mathbb{R}.$$

These remarks being made, we now mimick the development of Subsection **7.1** by considering, instead of the stochastic equation (7.a), Itô's formula applied to $\cosh(B_t)$, for instance; then, we obtain:

$$\cosh(B_t) = 1 + \int_0^t \sinh(B_s)dB_s + \frac{1}{2}\int_0^t \cosh(B_s)ds. \qquad (7.\text{m})$$

Following closely the method developed in **7.1**, we introduce:

$$S_t \overset{(\text{def})}{=} \int_0^t ds \, (\sinh(B_s))^2, \quad \text{and } \tau_t \equiv \inf\{u : S_u > t\}.$$

We then deduce from (7.k) that the process $(X_t \equiv \cosh(B_{\tau_t}),\ t \geq 0)$ satisfies:

$$X_t = 1 + \beta_t + \frac{1}{2}\int_0^t \frac{X_s \, ds}{(X_s^2 - 1)}, \qquad (7.\text{m}')$$

where $(\beta_t, t \geq 0)$ is a Brownian motion, and we have used again the relation: $\cosh^2(b) - \sinh^2(b) = 1$ to transform (7.m) into (7.m').

Remark that, by definition, the process $(X_t, t \geq 0)$ satisfies: $X_t \geq 1$, so that it may be more natural to write the equation (7.m') in terms of $\hat{X}_t \overset{(\text{def})}{=} X_t - 1$; we then obtain:

$$\hat{X}_t = \beta_t + \frac{1}{2}\int_0^t \frac{(\hat{X}_s + 1)ds}{(\hat{X}_s(2 + \hat{X}_s))}. \qquad (7.\text{m}'')$$

Likewise, we may replace the equation (7.a) by Itô's formula applied to $\sinh(B_t)$; then, we obtain:

$$\sinh(B_t) = \int_0^t \cosh(B_s)dB_s + \frac{1}{2}\int_0^t ds \, \sinh(B_s), \qquad (7.\text{n})$$

and introducing:

$$C_t = \int_0^t ds \, \cosh^2(B_s), \quad \text{and its inverse: } \sigma_t = \inf\{u : C_u > t\},$$

we see that the process: $Y_t = \sinh(B_{\sigma_t}),\ t \geq 0$, satisfies:

$$Y_t = \gamma_t + \frac{1}{2}\int_0^t \frac{Y_s \, ds}{1 + Y_s^2}, \qquad (7.\text{n}')$$

where $(\gamma_t, t \geq 0)$ is a one-dimensional Brownian motion, starting from 0. Here, $(Y_t, t \geq 0)$ takes values in $\mathbb{R}$, and is symmetric, i.e.:

$$(Y_t, t \geq 0) \overset{(law)}{=} (-Y_t, t \geq 0).$$

To push the method further, we now develop some of the computations related to $(C_t, t \geq 0)$, and therefore to the process $(Y_t, t \geq 0)$.

The process $(\sigma_t, t \geq 0)$ may be expressed explicitly in terms of $(Y_t, t \geq 0)$ as:

$$\sigma_t = \int_0^t \frac{ds}{1 + Y_s^2}, \quad t \geq 0,$$

and the rôle of the generator $\tilde{L}$, resp of φ_λ, in (7.1), is now played by the generator $\tilde{\mathcal{L}}$ of the process $(Y_t, t \geq 0)$, resp. by the function ψ_λ, both of which we now define:

$$\tilde{\mathcal{L}}\psi(y) = \frac{1}{2}\psi''(y) + \frac{1}{2}\left(\frac{y}{1+y^2}\right)\psi'(y)$$

and ψ_λ is a solution of:

$$\tilde{\mathcal{L}}\psi_\lambda(y) = \frac{\lambda\psi_\lambda(y)}{(1+y^2)}, \quad \text{with } \psi_\lambda(0) = 1$$

and ψ_λ is $\mathbb{R}_+$-valued.

Finally, if $Q_u^\lambda(z, dy) = q_u^\lambda(z, y)dy$ denotes the semi-group of the process with infinitesimal generator:

$$\tilde{\mathcal{L}}_\lambda = \tilde{\mathcal{L}} + \frac{\psi_\lambda'}{\psi_\lambda}\frac{d}{dy},$$

the analogue of formula (7.i) is:

$$P(\sinh(B_{T_\lambda}) \in dy; \; C_{T_\lambda} \in du) = \lambda q_u^\lambda(0, y)\frac{du\, dy}{(1+y^2)\psi_\lambda(y)}. \tag{7.o}$$

If one is only interested in the distribution of C_{T_λ}, there is the analogue of formula (7.e):

$$P(C_{T_\lambda} > u) = E_0^\lambda\left[\frac{1}{\psi_\lambda(Y_u)}\right]. \tag{7.p}$$

8.   Some Computational Issues

8.1. From the applied point of view, the formula

$$\lambda \int_0^\infty dt\, e^{-\lambda t}\, E[(A_t^{(\nu)} - k)^+] = \frac{\int_0^{1/2k} e^{-t}\, t^{\frac{\mu-\nu}{2}-2}(1 - 2kt)^{\frac{\mu+\nu}{2}+1}}{(\lambda - 2(1+\nu))\Gamma\left(\frac{\mu-\nu}{2} - 1\right)} \tag{1.1}$$

is not completely satisfactory; one would like to invert the Laplace transform in λ, which would give an expression of

$$E[(A_t^{(\nu)} - k)^+]. \tag{1.h}$$

In order to do this, we may divide both sides of (1.l) by λ, and then inspect the (new) right-hand side of (1.l), call it: $r(\lambda)$; we can write:

$$r(\lambda) = \frac{1}{\lambda(\lambda - 2(1+\nu))} \int_0^1 \frac{du}{u} e^{-u/2k} \left\{ \frac{\left(\frac{u}{2k}\right)^{\frac{\mu-\nu}{2}-1}}{\Gamma\left(\frac{\mu-\nu}{2}-1\right)} \right\} (1-u)^{\frac{\mu+\nu}{2}+1} \tag{8.a}$$

8.2. Another computational question, which also arises naturally in the applications of the results of the present paper to financial mathematics is that of finding an explicit expression for the distribution of:

$$H_t^{(\nu)} = \int_0^t \frac{du}{(R_u^{(\nu)})^2},$$

where $(R_u^{(\nu)}, u \geq 0)$ is a Bessel process with index ν, starting from 1, say.

This problem is closely related to the problem raised in **8.1**.

8.3. Coming back to **8.1**, we recall that, as indicated at the beginning of this paper, it is the expression:

$$E\left[\left(\frac{1}{t}A_t^{(\nu)} - k\right)^+\right] \tag{1.h'}$$

which is of interest, rather than (1.h).

References

1. Bougerol, Ph. (1983). Exemples de théorèmes locaux sur les groupes résolubles. *Ann. I.H.P.*, **19** (4), 369–391
2. Davis, B. (1979). Brownian motion and analytic functions. *Ann. Prob.*, **7**, 913–932
3. Dufresne, D. (1990). The distribution of a perpetuity, with applications to risk theory and pension funding. *Scand. Actuarial J.*, 39–79
4. Fuchs, A. and Letta, G. (1991). Un résultat élémentaire de fiabilité. Application à la formule de Weierstrass sur la fonction gamma. *Séminaire de Probabilités XXV. Lect. Notes in Maths. 1485.* p. 316–323, Springer
5. Geman, H. and Yor, M. (October 1993). Bessel processes, Asian options and perpetuities. *Mathematical Finance*, **3** (4), 349–375. **Paper [5] in this book**
6. Hartman, P. (1976). Completely monotone families of solutions of n-th order linear differential equations and infinitely divisible distributions. *Ann. Scuola Normale Superiore – Pisa – Classe di Scienze – Serie IV*, **III** (2), 267–287
7. Itô, K. and Mc Kean, H.P. (1965). *Diffusion processes and their sample paths.* Springer

8. Kawazu, K. and Tanaka, H. (1993). On the maximum of a diffusion process in a drifted Brownian environment. *Sém. Prob. XXVII, Lect. Notes in Maths. 1557,* Springer, p. 78–85

9. Lebedev, N.N. (1972). *Special functions and their applications,* Dover Publications

10. Lévy, P. (1940). Le mouvement brownien plan. *Amer. J. Math.,* **62**, 487–550

11. Patterson, S.J. (1988). An introduction to the theory of the Riemann Zeta function. *Cambridge Studies in Advanced Mathematics,* **14**

12. Pitman, J. and Yor, M. (1986). Level crossings of a Cauchy process. *Ann. Prob.,* **14** (3), 780–792

13. Pitman, J.W. and Yor, M. (1982). A decomposition of Bessel bridges. *Zeit. für Wahr.,* **59**, 425–457

14. Watson, G.N. (1966). *A treatise on the theory of Bessel functions.* Cambridge University Press

15. Yor, M. (1992). Sur certaines fonctionnelles exponentielles du mouvement brownien réel. *J. Appl. Proba.,* **29**, 202–208. **Paper [1] in this book**

16. Yor, M. (1992). On some exponential functionals of Brownian motion. *Adv. Appl. Prob.,* **24**, 509–531. **Paper [2] in this book**

17. Yor, M. (Juin 1992). Sur les lois des fonctionnelles exponentielles du mouvement brownien, considérées en certains instants aléatoires. *Note aux Comptes Rendus Acad. Sci. Paris, t. 314, Série I,* 951–956. **Paper [4] in this book**

18. Yor, M. (1992). *Some aspects of Brownian motion.* Part I: Some special functionals. *Lectures in Mathematics. ETH Zürich.* Birkhäuser

19. Black, F., Derman, E. and Toy, W. (January–February 1990). A one-factor model of interest rates and its application to Treasury Bond Options. *Financial Analysts Journal,* 33–39

20. Black, F. and Karasinski, P. (July–August 1991). Bond and Option pricing when short rates are lognormal. *Financial Analysts Journal,* 52–59

21. Hull, J. and White, A. (1990). Pricing interest rate derivative securities. *Review of Financial Studies,* **3**, 573–592

22. Dothan, L.U. (1978). On the term structure of interest rates. *Journal of Financial Economics,* **6**, 59–69

23. Gordon, L. (1994). A stochastic approach to the Gamma function. *Amer. Math. Monthly,* **101**, 858–865

Postscript #6

The present paper is being published for the first time. Its main aim is to discuss analogous quantities to those being studied throughout this book for geometric Brownian motion, which here is being replaced by some diffusions.

As this book is going to press, the computational issues raised in Section 8 are still being actively studied by, e.g: V. Linetsky, P. Carr, M. Schröder, …

From Planar Brownian Windings
to Asian Options

Insurance: Mathematics and Economics **13** (1993), 23–34

Abstract. It is shown how results presented in *Insurance: Mathematics and Economics* 11, no. 4, in several papers by De Schepper, Goovaerts, Delbaen and Kaas, concerning the arithmetic average of the exponential of Brownian motion with drift [which plays an essential role in Asian options, and has also been studied by the author, jointly with H. Geman] are related to computations about winding numbers of planar Brownian motion. Furthermore, in the present paper, Brownian excursion theory is being used in an essential way, and helps to clarify the role of some Bessel functions computations in several formulae.

1. Introduction

1.1. The main purpose of this paper is to explain the deep relations which exist between the distribution of the (continuous determination of the) winding number θ_t, taken at some time $t > 0$, of planar Brownian motion $(Z_u, u \geq 0)$, with $Z_0 = z_0 \neq 0$, around 0, and the distribution of

$$A_t^{(\nu)} \stackrel{\text{def}}{=} \int_0^t ds \exp 2(B_s + \nu s),$$

where $(B_s, s \geq 0)$ denotes a linear Brownian motion, starting from 0, and $\nu \in \mathbb{R}$.

The price of (financial) Asian options is the quantity

$$C^{(\nu)}(t; k) \stackrel{\text{def}}{=} E\left[\left(\frac{1}{t} A_t^{(\nu)} - k\right)^+\right],$$

and a discussion and analysis of this quantity, and related ones, from a mathematical finance point of view is done in Geman and Yor (1993).

1.2. The Ariadne thread which runs between the (computation of the) distribution of θ_t on one hand, and that of $A_t^{(\nu)}$ on the other hand, is not so easy to unravel, in particular because Bessel processes, and/or Bessel functions crop up very naturally, at least as intermediate objects, and sometimes as essential ingredients. To illustrate this, we present three formulae:

(i) In order to recover Spitzer's asymptotic result [Spitzer (1958)],

$$\frac{2}{\log t}\theta_t \xrightarrow[t\to\infty]{(\text{law})} C_1, \tag{1}$$

where C_1 denotes a Cauchy variable with parameter 1, Itô and McKean (1965, p. 271, formulae (12), (13)) start by proving the formulae[1]

$$\mathbb{E}_{z_0}\left[\int_0^\infty dt \exp\left(-\frac{\beta^2 t}{2} + i\mu\theta_t\right)\right] = \int_0^\infty b\, db\, G_{|\mu|}(\beta a, \beta b) \tag{2}$$

$$= \int_0^\infty dt \exp\left(-\frac{\beta^2 t}{2}\right)\int_0^\infty \frac{b\, db}{t}\exp\left(-\frac{a^2+b^2}{2t}\right)I_{|\mu|}\left(\frac{ab}{t}\right), \tag{3}$$

where $a = |z_0|$ and the notation $G_\mu(u,v)$, used in (2), means

$$G_\mu(u,v) = \begin{cases} 2I_\mu(u)K_\mu(v), & \text{if } u \le v \\ 2K_\mu(u)I_\mu(v), & \text{if } v \le u \end{cases}.$$

(ii) In De Schepper et al. (1992), one of the main results of the authors is the formula

$$\int_0^\infty dt \exp(-st)E\left[\exp-\frac{u^2}{2}A_t^{(\nu)}\right] = \int_{-\infty}^\infty dy \exp(\nu y)G_\mu(u, u\, e^y), \tag{4}$$

where $\mu = \sqrt{2s + \nu^2}$.

Our main incentive to write this paper was our desire to explain the relationship between formulae (2) and (4).

(iii) Obviously, if we multiply both sides of (4) by s, we obtain an expression of

$$E\left[\exp\left(-\frac{u^2}{2}A_{T_s}^{(\nu)}\right)\right],$$

where T_s denotes an exponential variable, with parameter s, which is independent of $(B_t, t \ge 0)$.

The distribution of $A_{T_s}^{(\nu)}$ has been obtained in Yor (1992c), and is simple enough; indeed, one has

$$A_{T_s}^{(\nu)} \stackrel{(\text{law})}{=} \frac{1 - U^{1/a}}{2Z_b}, \tag{5}$$

where U is a uniform r.v. on $[0,1]$, and Z_b is a gamma variable with parameter b, independent of U; moreover, $a = (\mu + \nu)/2$, $b = (\mu - \nu)/2$, and $\mu = \sqrt{2s + \nu^2}$. Consequently:

$$P(A_{T_s}^{(\nu)} \ge u) = \frac{1}{\Gamma(b)}\int_0^{1/2u} dv \exp(-v)v^{b-1}(1 - 2uv)^a \tag{6}$$

---

[1] $\mathbb{E}_{z_0}$ denotes the expectation operator associated with the law of $(Z_t, t \ge 0)$, when $Z_0 = z_0$.

and

$$P(A_{T_s}^{(\nu)} \in du) = \frac{(2a)du}{\Gamma(b)} \int_0^{1/2u} dv \exp(-v)v^b(1-2uv)^{a-1}, \tag{7}$$

$$E[(A_{T_s}^{(\nu)})^\gamma] = \frac{\Gamma(1+\gamma)\Gamma(((\mu+\nu)/2)+1)\Gamma(((\mu-\nu)/2)-\gamma)}{2^\gamma\Gamma((\mu-\nu)/2)\Gamma(1+\gamma+((\mu+\nu)/2))}, s > 2\gamma(\gamma+\nu). \tag{8}$$

1.3. The great variety of backgrounds (e.g.: theoretical physics, classical analysis, semigroup theory, and, of course, mathematical finance) of the scientists who have been, or still are, interested in one of the above quantities makes it more than likely that such formulae will be discovered again and again; it is our hope that the present exposition may put the different approaches considered so far in perspective.

1.4. In order to emphasize the importance of time-changes, changes of probabilities, excursion theory, used throughout this paper, the details of differential equations computations will not be presented here. I am quite confident that every differential equation aficionado will easily fill in the gaps. However, the following precision may be useful: although Bessel functions appear in many places throughout this paper, very little knowledge about them is necessary.

In fact, their appearance in the present paper, and in the other papers [Yor (1992a, b, c)] written by the author on related subjects, can simply be traced back to the fact that, if $P_a^{(\nu)}$ denotes the law of the Bessel process $(R_t, t \geq 0)$ with dimension $d_\nu \equiv 2(\nu+1)$, starting at $a > 0$, one has

$$E_a^{(\nu)}[\exp(-sR_t^2)] = \frac{1}{(1+2st)^{1+\nu}} \exp\left(-\frac{a^2s}{1+2st}\right). \tag{9}$$

From this Laplace transform, the semigroup $p_t^\nu(a,r)dr$ of the Bessel process with dimension d_ν is easily found to be given by the formula

$$p_t^{(\nu)}(a,r) = \frac{1}{t}\left(\frac{r}{a}\right)^\nu r \exp\left(-\frac{(a^2+r^2)}{2t}\right) I_\nu\left(\frac{ar}{t}\right), \text{ for } t > 0, a, r > 0. \tag{10}$$

1.5. We now end this introduction with an outline of the plan of the remaining sections of the paper.

In Section 2, a number of fundamental relations between planar Brownian motion, Bessel processes, and exponentials of linear Brownian motion, are presented. Section 2 also contains a short, but hopefully useful, presentation of some master formulae in Brownian excursion theory.

Section 3 consists in a quick derivation of the law of $A_{T_s}^{(\nu)}$, which is presented above in formulae (5), (6), (7), (8). In Section 4, we show how formula (4) is closely related with formulae (2) and (3), thanks to the time-changes presented in Section 2. Some further results about the law of the process $(A_t^{(\nu)}, t \geq 0)$ taken at various random times are presented.

Finally, in Section 5, it is shown how the distributional results obtained in Section 4 for the process $(A_t^{(\nu)}, t \geq 0)$ may be translated into distributional results about the winding number of planar Brownian motion.

2. The Ariadne Thread, i.e., Some Relations Between Planar Brownian Motion, Bessel Processes, and Exponentials of Linear Brownian Motion

The material presented in this section may be found partly in Yor (1992a); here, some particular emphasis is put on planar Brownian motion and, in Subsection **2.5**, on master formulae in excursion theory.

2.1.   The Skew-product Representation of Planar Brownian Motion

If $(Z_t, t \geq 0)$ denotes planar Brownian motion, starting from $z_0 \neq 0$, it can be represented as

$$Z_t = R_t \, \exp(i\theta_t) \equiv R_t \exp(i\beta_{H_t}), \tag{11}$$

where $R_t = |Z_t|, \{\theta_t, t \geq 0\}$ denotes the winding number of $(Z_t, t \geq 0)$ around $0, H_t = \int_0^t (ds/R_s^2)$, and $(\beta_u, u \geq 0)$ is a one-dimensional Brownian motion independent of $(R_t, t \geqslant 0)$.

Here is a simple, but interesting, consequence of the representation (11): for every $\nu \in \mathbb{R}$, and every Borel function $f : \mathbb{R}^+ \to \mathbb{R}^+$, one has

$$E[f(R_t) \exp(i\nu\theta_t)] = E\left[f(R_t) \exp\left(-\frac{\nu^2}{2} H_t\right)\right]. \tag{12}$$

2.2.   The Cameron–Martin Relationship

If F is an $\mathbb{R}^+$-valued functional defined on $C(\mathbb{R}^+, \mathbb{R})$, we have, for every $t \geq 0$, and $\nu \in \mathbb{R}$,

$$E[F(B_s + \nu s; \ s \leq t)] = E\left[F(B_s; s \leq t) \exp\left(\nu B_t - \frac{\nu^2 t}{2}\right)\right].$$

2.3.   Some Time-changes

For simplicity, we assume $\nu \geq 0$; define $B_t^{(\nu)} = B_t + \nu t$. Then, we have

$$\exp(B_t^{(\nu)}) = R_{A_t^{(\nu)}}^{(\nu)}, \quad \text{where} \quad A_t^{(\nu)} \stackrel{\text{def}}{=} \int_0^t ds \exp(2B_s^{(\nu)}),$$

and $(R_u^{(\nu)}, u \geq 0)$ is a Bessel process, with index ν, starting from 1. Conversely, using the same notation, we have

$$\log(R_u^{(\nu)}) = B_s^{(\nu)}\big|_{s \equiv H_u^{(\nu)} = \int_0^u (dv/(R_v^{(\nu)})^2)}.$$

2.4. The Girsanov Relationship Between the Laws of Two Different Bessel Processes

Here, we consider $\nu \geq 0$, and $P^{(\nu)}$ denotes the law of the Bessel process with index ν, starting from 1. This law is defined on $C(\mathbb{R}^+, \mathbb{R}^+)$, with (R_t) the coordinate process, and $\mathscr{R}_t = \sigma\{R_s; s \leq t\}$. Then, we have

$$P^{(\nu)}_{|\mathscr{R}_t} = R_t^\nu \exp\left(-\frac{\nu^2}{2}\int_0^t \frac{ds}{R_s^2}\right) \cdot P^{(0)}_{|\mathscr{R}_t}.$$

This result may be understood as a consequence of the Cameron–Martin relationship recalled in **2.2**, after time-changing according to **2.3**.

Putting together the absolute continuity relationship between $P^{(\nu)}$ and $P^{(0)}$, and the explicit formula (10) for $p_t^{(\nu)}(a, r)$, the following conditional Laplace transform formula ensues:

$$E_a^{(0)}\left(\exp -\frac{\nu^2}{2} H_t \mid R_t = r\right) = \frac{I_{|\nu|}}{I_0}\left(\frac{ar}{t}\right).$$

2.5. A Master Formula in Excursion Theory

For simplicity, we shall only consider here $\mathbb{R}^+$-valued continuous additive functionals of the Brownian motion $(B_t; t \geq 0)$, which are of the form

$$F_t = \int_0^t ds\, f(B_s), \text{ for some Borel function } f : \mathbb{R} \to \mathbb{R}^+.$$

In this section, S_μ denotes an exponential variable with parameter $\mu^2/2$, which is independent of $(B_t; t \geq 0)$; moreover, let $g_t = \sup\{s \leq t; B_s = 0\}$ $(t \geq 0)$. Then, we have:

(i) the variables

$$F_{g_{S_\mu}} \text{ and } (F_{S_\mu} - F_{g_{S_\mu}})$$

are independent, and satisfy the formulae

$$E[\exp(-F_{g_{S_\mu}})] = \mu \int_0^\infty ds E\left[\exp -\left(F_{\tau_s} + \frac{\mu^2}{2}\tau_s\right)\right], \tag{13}$$

where $(\tau_s, s \geq 0)$ denotes the inverse of the local time $(l_u; u \geq 0)$ at 0 of $(B_u; u \geq 0)$;

$$E[\exp -(F_{S_\mu} - F_{g_{S_\mu}})] = \frac{\mu}{2}\int_{-\infty}^\infty da E_a\left[\exp -\left(F_{T_0} + \frac{\mu^2}{2}T_0\right)\right], \tag{14}$$

where $T_0 \equiv \inf\{t \geq 0 : B_t = 0\}$.

As a consequence, $E[\exp(-F_{S_\mu})]$ is equal to the product of the expressions on the right-hand sides of (13) and (14).

(ii) To be more complete, the processes

$$(B_u; u \leq g_{S_\mu}) \quad \text{and} \quad (B_{g_{S_\mu}+v}; v \leq S_\mu - g_{S_\mu})$$

are independent; as a consequence, the pairs

$$\{F_{g_{S_\mu}}; l_{S_\mu}\} \quad \text{and} \quad \{F_{S_\mu} - F_{g_{S_\mu}}; B_{S_\mu}\}$$

are independent, and satisfy

(ii′) $P(l_{S_\mu} \in \mathrm{d}s) = \mu\exp(-\mu s)\,\mathrm{d}s$ and $P(B_{S_\mu}\in \mathrm{d}a) = \dfrac{\mu}{2}e^{-\mu|a|}\,\mathrm{d}a.$

The formulae (13) and (14) can be further disintegrated into

$$E[\exp(-F_{g_{S_\mu}})|\,l_{S_\mu} = s] = E\left[\exp-\left(F_{\tau_s} + \frac{\mu^2}{2}\tau_s\right)\right]e^{\mu s}, \qquad (13')$$

$$E[\exp-(F_{S_\mu} - F_{g_{S_\mu}})|B_{S_\mu} = a] = E_a\left[\exp-\left(F_{T_0} + \frac{\mu^2}{2}T_0\right)\right]e^{\mu|a|} \qquad (14')$$

As a conclusion of this subsection, we remark that these formulae give an excursion theoretic interpretation of Feynman–Kac formulae; this approach has already been used in a number of examples: see, e.g., the extensions by Petit (1992) of Paul Lévy's arc sine distribution for Brownian motion [developed also in Yor (1992b, Chapters 8, 9)] or the Ray–Knight theorem for Brownian local times taken at an independent exponential time [Biane and Yor (1988)].

A main part of the present paper–Section 4 below–consists in the application of the above formulae to the case $f(x) = \exp(\alpha x)$, or $\exp(\alpha|x|)$. Some (perhaps simpler) applications of these formulae to certain extensions of the arc sine law for Brownian motion, discovered by Paul Lévy in 1939, are found in Barlow et al. (1989), and also in Yor (1992b).

3. The Distribution of $A_{T_s}^{(\nu)}$

In order to obtain the law of $A_{T_s}^{(\nu)}$, as described by formula (5), we start with the following chain of equalities, which is a consequence of the time-change relations discussed in Subsection **2.3**:

$$P(A_{T_s}^{(\nu)} \geq u) = s\int_0^\infty \mathrm{d}t\,\exp(-st)P(A_t^{(\nu)} \geq u)$$

$$= s\int_0^\infty \mathrm{d}t\,\exp(-st)P(H_u^{(\nu)} \leq t) = E[\exp(-sH_u^{(\nu)})],$$

where

$$H_u^{(\nu)} \overset{\text{def}}{=} \int_0^u \frac{dh}{(R_h^{(\nu)})^2}.$$

Using now the absolute continuity relationship presented in Subsection **2.4**, we obtain

$$E[\exp(-sH_u^{(\nu)})] = E^{(0)}\left[(R_u)^\nu \exp\left(-\frac{\mu^2}{2}H_u\right)\right] = E^{(\mu)}\left[\frac{1}{(R_u)^{\mu-\nu}}\right],$$

where $\mu = \sqrt{2s + \nu^2}$.

Hence, we have obtained the equality

$$P(A_{T_s}^{(\nu)} \geq u) = E^{(\mu)}\left[\frac{1}{(R_u)^{\mu-\nu}}\right],$$

and, from formula (9), we deduce

$$E^{(\mu)}\left(\frac{1}{(R_u)^{2b}}\right) = \frac{1}{\Gamma(b)}\int_0^{1/2u} dv \exp(-v)v^{b-1}(1 - 2uv)^a$$

with $a = (\mu + \nu)/2$, and $b = (\mu - \nu)/2$.

Hence, we have now proved formula (6), which gives a closed form formula for the tail of the distribution of $A_{T_s}^{(\nu)}$. The representation of $A_{T_s}^{(\nu)}$ as the beta-gamma ratio presented in (5) is then easily obtained.

4. Decomposing A_{S_μ}

4.1. The aim of this section is to provide a better understanding of formula (4). We first remark that, thanks to the Cameron–Martin relationship between the laws of $(B_s^{(\nu)}; s \leq t)$ and $(B_s; s \leq t)$, formula (4) may be written as

$$E\left[\exp\left(-\frac{u^2}{2}A_{S_\mu} + \nu B_{S_\mu}\right)\right] = \frac{\mu^2}{2}\int_{-\infty}^\infty dy \exp(\nu y)G_\mu(u, u\ e^y). \tag{15}$$

As recalled in Subsection **2.5**, we have $P(B_{S_\mu} \in dy) = (\mu/2)\exp(-\mu|y|)dy$; hence, formula (15) may be disintegrated into

$$E\left[\exp\left(-\frac{u^2}{2}A_{S_\mu}\right)\Big| B_{S_\mu} = y\right] = \mu\exp(\mu|y|)G_\mu(u, u\ e^y), \tag{16}$$

which, thanks to the end of the discussion in **2.5**, may be further decomposed into

$$E\left[\exp\left(-\frac{u^2}{2}A_{g_{S_\mu}}\right)\right] = \mu G_\mu(u, u) \equiv 2\mu I_\mu(u)K_\mu(u), \tag{17}$$

and

$$E\left[\exp-\frac{u^2}{2}(A_{S_\mu}-A_{g_{S_\mu}})|B_{S_\mu}=y\right]=\mu\exp(\mu|y|)\frac{G_\mu(u,u\,e^y)}{G_\mu(u,u)}$$

$$\equiv\begin{cases}\mu\exp(\mu|y|)\dfrac{K_\mu(u\,e^y)}{K_\mu(u)} & \text{if}\quad y\geq 0\\[3mm]\mu\exp(\mu|y|)\dfrac{I_\mu(ue^y)}{I_\mu(u)} & \text{if}\quad y\leq 0\end{cases}\tag{18}$$

In order to gain a better understanding of these formulae, we present [and leave the proof to the reader; see Subsection **1.4** above!] the following

Lemma 1. *Define*

$$A_t^\pm\equiv\int_0^t ds\exp(2B_s)1_{(B_s\in\mathbb{R}^\pm)},\quad \tau_s^\pm=\int_0^{\tau_s}du1_{(B_u\in\mathbb{R}^\pm)},$$

where $\tau_s\equiv\inf\{u:l_u>s\}$, and $(l_u,u\geq 0)$ is the local time at 0 of B. Furthermore, let P_x denote the law of $(B_u,u\geq 0)$, with $B_0=x$, and define $T_0\equiv\inf\{t\geq 0:B_t=0\}$. Then, we have, for every $m,n,\mu,\nu\geq 0$, and $x\in\mathbb{R}$, the formulae:

(a) $$E_0[\exp-\tfrac{1}{2}(m^2A_{\tau_s}^++\mu^2\tau_s^+)]=\exp-\tfrac{s}{2}\left(\frac{mK_{\mu+1}(m)}{K_\mu(m)}-\mu\right),$$

(b) $$E_0[\exp-\tfrac{1}{2}(n^2A_{\tau_s}^-+\mu^2\tau_s^-)]=\exp-\tfrac{s}{2}\left(\frac{nI_{\mu-1}(n)}{I_\mu(n)}-\mu\right),$$

(c) $$E_x[\exp-\tfrac{1}{2}(m^2A_{T_0}+\nu^2T_0)]=\frac{I_\nu(m\,e^{-x})}{I_\nu(m)}\quad(x\geq 0),$$

(d) $$E_x[\exp-\tfrac{1}{2}(m^2A_{T_0}+\nu^2T_0)]=\frac{K_\nu(m\,e^{-x})}{K_\nu(m)}\quad(x\leq 0).$$

We now show how the formulae presented in Lemma 1 allow to check formulae (17) and (18), and to gain more insight.

For example, in order to check formula (17), we remark that we have, thanks to excursion theory,

$$E[\exp-\tfrac{1}{2}(m^2A_{g_{S_\mu}}^++n^2A_{g_{S_\mu}}^-)]$$

$$=\mu\int_0^\infty ds E\left[\exp-\left(\frac{m^2}{2}A_{\tau_s}^++\frac{\mu^2}{2}\tau_s^+\right)\right]E\left[\exp-\left(\frac{n^2}{2}A_{\tau_s}^-+\frac{\mu^2}{2}\tau_s^-\right)\right]\tag{19}$$

$$=2\mu\left\{\frac{mK_{\mu+1}(m)}{K_\mu(m)}+\frac{nI_{\mu-1}(n)}{I_\mu(n)}-2\mu\right\}^{-1},$$

so that, taking $m = n = u$, we obtain, after a little algebra,

$$E\left[\exp\left(-\frac{u^2}{2}A_{g_{S_\mu}}\right)\right] = 2\mu I_\mu(u)K_\mu(u), \tag{17'}$$

which is nothing else but formula (17).

To go from (19), with $m = n = u$, to (17'), we have used the Wronskian relation,

$$W\{I_\mu(z), K_\mu(z)\} \equiv I'_\mu(z)K_\mu(z) - I_\mu(z)K'_\mu(z) = \frac{1}{z},$$

together with the relations

$$K_{\mu+1}(z) = -K'_\mu(z) + \frac{\mu}{z}K_\mu(z) \quad\text{and}\quad I_{\mu-1}(z) = I'_\mu(z) + \frac{\mu}{z}I_\mu(z)$$

[for these relations, a very convenient reference is Lebedev (1972, pp. 110, 113)] in order to prove

$$\left\{\frac{uK_{\mu+1}(u)}{K_\mu(u)} + \frac{uI_{\mu-1}(u)}{I_\mu(u)} - 2\mu\right\}^{-1} = I_\mu(u)K_\mu(u).$$

Now, (17') clearly appears as a consequence of (19).

4.2. In the preceding subsection, we have developed computations not only for $A_t \equiv \int_0^t ds\,\exp(2B_s)$, but also for (A_t^+) and (A_t^-). This gives us, at no extra cost, some results for the following additive functionals:

$$A'_t \equiv \int_0^t ds\,\exp(2|B_s|) \quad\text{and}\quad A''_t = \int_0^t ds\,\exp(-2|B_s|).$$

Indeed, from the formulae (a) and (b) in Lemma 1, we deduce

$$E\left[\exp-\frac{1}{2}\left(m^2\int_0^{T_s} du\,\exp(2|B_u|) + \mu^2\tau_s\right)\right]$$
$$= \exp-s\left(\frac{mK_{\mu+1}(m)}{K_\mu(m)} - \mu\right),$$

$$E\left[\exp-\frac{1}{2}\left(m^2\int_0^{T_s} du\,\exp-(2|B_u|) + \mu^2\tau_s\right)\right]$$
$$= \exp-s\left(\frac{mI_{\mu-1}(m)}{I_\mu(m)} - \mu\right).$$

Now, from the recurrence relations between Bessel functions, we deduce, in particular [see Lebedev (1972, p. 110)],

$$\frac{xI_{\mu-1}(x)}{I_\mu(x)} - \mu = \frac{xI_{\mu+1}(x)}{I_\mu(x)} + \mu, \quad \frac{xK_{\mu+1}(x)}{K_\mu(x)} - \mu = \frac{xK_{\mu-1}(x)}{K_\mu(x)} + \mu$$

and, consequently, we obtain

$$E\left[\exp\left(-\frac{m^2}{2}\int_0^{g_{S_\mu}} du\, \exp(2|B_u|)\right)\right] = \frac{\mu}{(mK_{\mu+1}(m)/K_\mu(m)) - \mu}$$

$$= \frac{\mu}{(mK_{\mu-1}(m)/K_\mu(m)) + \mu},$$

$$E\left[\exp\left(-\frac{m^2}{2}\int_0^{g_{S_\mu}} du\, \exp(-2|B_u|)\right)\right] = \frac{\mu}{(mI_{\mu-1}(m)/I_\mu(m)) - \mu}$$

$$= \frac{\mu}{(mI_{\mu+1}(m)/I_\mu(m)) + \mu}.$$

4.3. Using again formulae (a) and (b) in Lemma 1, we obtain some even simpler formulae for the Laplace transforms of the laws of $A^\pm_{g_{S_\mu}}$; precisely, we find

$$E\left[\exp\left(-\frac{m^2}{2}A^+_{g_{S_\mu}}\right)\right] = \frac{2\mu K_\mu(m)}{mK_{\mu+1}(m)}, \tag{20}_+$$

and

$$E\left[\exp\left(-\frac{m^2}{2}A^-_{g_{S_\mu}}\right)\right] = \frac{2\mu I_\mu(m)}{mI_{\mu-1}(m)}. \tag{20}_-$$

The right-hand sides of the formulae given in Lemma 1, and of formulae (19) and $(20)_\pm$ are also found in Getoor and Sharpe (1979) and in Pitman and Yor (1981) in some computations for Bessel processes.

For example, we find, from the injectivity of the Laplace transform, that, comparing formula $(20)_-$ and formula (9.h) [see, in fact, more generally, Proposition (9.2)] in Pitman and Yor (1981), one has

$$A_{g_{S_\mu}} \stackrel{\text{(law)}}{=} \int_0^\infty ds1_{(R_s^{(\mu)} \leq 1)}, \tag{$*$}$$

where $(R_s^{(\mu)}, s \geq 0)$ denotes here a Bessel process with index $\mu > 0$, starting from 1. The interested reader, perusing through Pitman and Yor (1981), will easily find several other instances of such identities in law, which may all be explained by the following

Lemma 2. [Williams (1974)]. *Let $\mu > 0$, and let S_μ denote an exponential variable with parameter $\mu^2/2$, independent of $(B_u, u \geq 0)$. Then, the process $(B_u, u \leq S_\mu)$, conditioned on $B_{S_\mu} = x \geq 0$ is identical in law to*

$$(B_u + \mu u;\ u \leq L_x^\mu \stackrel{\text{def}}{=} \sup\{t : B_t + \mu t = x\}).$$

In particular, for $x = 0$, one has

$$(B_u;\ u \leq g_{S_\mu}) \stackrel{\text{(law)}}{=} (B_u + \mu u;\ u \leq L_0^\mu).$$

Proof. As a particular case of application of Exercise (1.16) in Revuz and Yor (1991, p. 378), the law of $(B_u + \mu u; u \le L_x^u)$, given $L_x^\mu = t$, is that of the Brownian bridge of length t, with starting position 0, and ending position x. [The fact that this distribution does not depend on μ is a consequence of the Cameron–Martin relationship in Subsection **2.2**].

Moreover, it is easily shown that the law of S_μ, given $B_{S_\mu} = x$, is that of L_x^μ [again, Exercise (4.16), in Revuz and Yor (1991, p. 298), may be helpful].

The proof of Lemma 2 is then obtained by putting the two previous arguments together. □

In order to deduce the identity $(*)$ from Lemma 2, it now remains to remark, from Subsection **2.3**, that the right-hand side of $(*)$ is equal, in law, to

$$\int_0^\infty dt \, \exp(2B_t^{(\mu)}) 1_{(B_t^{(\mu)} \le 0)},$$

and $(*)$ now follows from the last assertion in Lemma 2.

As a last comment concerning formula $(20)_-$, it may be interesting to point out that the distribution of $A_{g_{S_\mu}}^-$ is, in the particular case $\mu = 3/2$, a theta function; more generally, one has

$$\frac{1}{m} \frac{I_\mu(m)}{I_{\mu-1}(m)} = 2 \sum_{k=1}^{\infty} \frac{1}{m^2 + j_{\mu-1,k}^2},$$

where $(j_{\nu,k}; k \ge 1)$ denotes the increasing sequence of the simple, positive, zeros of the Bessel function J_ν.

As a consequence, we obtain

$$P(A_{g_{S_\mu}}^- \in da) = \tfrac{1}{2}\theta^\mu\left(\tfrac{a}{2}\right) da, \quad \text{where} \quad \theta^\mu(t) = (4\mu) \sum_{k=1}^{\infty} \exp(-j_{\mu-1,k}^2 t).$$

In the particular case $\mu = 3/2$, we have $j_{\mu-1,k} \equiv k\pi$, from which we deduce the following probabilistic representation:

$$3 \cdot 2^{s/2} \frac{\Gamma\left(\tfrac{s}{2}\right)}{\pi^s} \zeta_R(s) = E[(A_{g_{S_{3/2}}}^-)^{s/2-1}]$$

of the Riemann zeta function $\zeta_R(s) \equiv \sum_{n=1}^\infty 1/n^s$ [for more developments, see Yor (1997, Chapter 11)].[1]

---

[1] See also *Ph. Biane, J. Pitman, M. Yor*: Probability laws related to the Riemann zeta and Jacobi theta functions, via Brownian excursions. To appear in Bull. AMS (2001).

4.4. Let $\nu' = -\nu$. Making use of the times changes described in (2.3.), we now obtain another expression, which we will compare later with formulae (4) and (15), for

$$
\begin{aligned}
\Phi_{\mu,\nu'}(u) &\overset{\text{def}}{=} \int_0^\infty dt \exp\left(-\frac{\mu^2 t}{2}\right) E\left[\exp\left(-\frac{u^2}{2}A_t + \nu' B_t\right)\right] \\
&= E_1^{(0)}\left[\int_0^\infty ds (R_s)^{\nu'-2} \exp -\tfrac{1}{2}(u^2 s + \mu^2 H_s)\right] \\
&= E_1^{(0)}\left[\int_0^\infty ds (R_s)^{\nu'-2} \exp\left(-\frac{u^2 s}{2}\right) E_1^{(0)}\left\{\exp\left(-\frac{\mu^2}{2}H_s\right) | R_s\right\}\right].
\end{aligned} \tag{21}
$$

Since, from the end of Subsection **2.4**, we know that

$$
E_1^{(0)}\left[\exp\left(-\frac{\mu^2}{2}H_s\right) | R_s\right] = \left(\frac{I_{|\mu|}}{I_0}\right)\left(\frac{R_s}{s}\right),
$$

we obtain

$$
\Phi_{\mu,\nu'}(u) = E_1^{(0)}\left[\int_0^\infty ds (R_s)^{\nu'-2} \exp\left(-\frac{u^2 s}{2}\right)\left(\frac{I_{|\mu|}}{I_0}\right)\left(\frac{R_s}{s}\right)\right].
$$

Using the explicit formula (10) of the density $p_s^{(0)}$ of the semi-group of the two-dimensional Bessel process, we obtain

$$
\begin{aligned}
&\Phi_{\mu,\nu'}(u) \\
&= \int_0^\infty ds \exp\left(-\frac{u^2 s}{2}\right) \int_0^\infty d\rho\, \rho^{\nu'-1}\left(\frac{1}{s}\right)\exp\left(-\frac{1+\rho^2}{2s}\right) I_\mu\left(\frac{\rho}{s}\right).
\end{aligned} \tag{22}
$$

We now pause to remark that, in the case $\nu' = 2$, the right-hand side of (22) is precisely formula (2) above, taken from Itô and McKean (1965); an explanation is that $\Phi_{\mu,\nu'}(u)$ is equal to formula (21), which, in turn, from the skew-product representation of planar Brownian motion is equal to the quantity in (2) above, related to the winding number of planar Brownian motion.

This remark being made, we now compare formulae (22) and (4), or (15). After making the change of variables $\rho = \exp(x)$ in (22), we find that both formulae are Laplace transforms with respect to $\nu = \nu'$. From the injectivity of the Laplace transform, we now deduce the equality

$$
\begin{aligned}
&G_\mu(u, u\exp(y)) \\
&= \int_0^\infty ds \exp\left(-\frac{u^2 s}{2}\right)\frac{1}{s}\exp\left(-\frac{1+\exp(2y)}{s}\right) I_\mu\left(\frac{\exp(y)}{s}\right)
\end{aligned}
$$

or, equivalently, putting $a = \exp(y)$,

$$G_\mu(u, ua) = \int_0^\infty \frac{ds}{s} \exp\left(-\frac{u^2 s}{2}\right) \exp\left(-\left(\frac{1+a^2}{2s}\right)\right) I_\mu\left(\frac{a}{s}\right). \qquad (23)$$

Formula (23) is, obviously, a purely analytical, and classical, formula which relates Bessel functions, and we feel that this may be a good place to end our circumnavigation around formulae (2), (3) and (4): indeed, we have just shown that formula (23) may be obtained by using the time changes in subsection (2.3.), and either of the formulae (3) or (4), whereas in De Schepper et al. (1992) or Itô and McKean (1965), the authors take (23) from a reference book on special functions, in order to deduce (3) or (4) from Feynman–Kac considerations.

5. Decomposing θ_{S_m}, the Winding Number of Planar Brownian Motion, taken at an Independent Exponential Time S_m, with Parameter $m^2/2$

5.1. *Some notation*: In this section, we consider complex Brownian motion $(Z_t, t \geq 0)$ starting from $z_0 = 1$. We define

$$\log R_t = B_{H_t} \quad \text{with} \quad H_t = \int_0^t \frac{ds}{R_s^2}; \ \theta_t^- = \int_0^t d\theta_s 1_{(R_s \leq 1)}; \ \theta_t^+ = \int_0^t d\theta_s 1_{(R_s \geq 1)}.$$

We let $\gamma_t = \sup\{s < t: R_s = 1\} \equiv g_{H_t}$, where $g_u \equiv \sup\{s < u: B_s = 0\}$, and we define (L_t) to be the local time of $(\log R_t; t \geq 0)$ at 0; hence, we have: $L_t = l_{H_t}$, where $(l_u, u \geq 0)$ denotes the local time at 0 for $(B_u, u \geq 0)$.

5.2. Our aim in this subsection is to translate the results obtained in Section 4 into results concerning θ_{S_m}, or, more precisely, the joint law of

$$\{\theta_{(m)}^- \overset{\text{def}}{=} \theta_{\gamma_{S_m}}^- ; \theta_{(m)}^+ \overset{\text{def}}{=} \theta_{\gamma_{S_m}}^+ ; \tilde{\theta}_{(m)} \overset{\text{def}}{=} \theta_{S_m} - \theta_{\gamma_{S_m}} ; L_{S_m} ; R_{S_m}\}.$$

The time-change arguments presented in Subsection **2.3**, and used in Subsection **4.3** apply here to give the multidimensional Laplace–Fourier transform

$$E[\exp\{i(\lambda\theta_{(m)}^- + \mu\theta_{(m)}^+ + \nu\tilde{\theta}_{(m)}) - aL_{S_m} + i\xi \log R_{S_m}\}],$$

which, hence, is equal to

$$\frac{m^2}{2} E\left[\int_0^\infty du \exp\left\{-\frac{m^2}{2} A_u + (i\xi + 2)B_u - al_u\right\} \cdots \right.$$

$$\left. \cdots \exp{-\tfrac{1}{2}\{\lambda^2 g_u^- + \mu^2 g_u^+ + \nu^2(u - g_u)\}}\right]$$

and this expression is, in turn, thanks to the master formula presented in Subsection **2.5**, equal to

$$\frac{m^2}{2} \int_0^\infty ds \exp(-as) E[\exp -\tfrac{1}{2}(m^2 A_{\tau_s} + \lambda^2 \tau_s^- + \mu^2 \tau_s^+)] \dots$$

$$\dots \int_{-\infty}^\infty dx \exp((i\xi + 2)x) E_x[\exp -\tfrac{1}{2}(m^2 A_{T_0} + \nu^2 T_0)].$$

Finally, using the four formulae presented in Lemma 1 in Subsection **4.1**, we obtain the following formula for the present five-parameter Laplace–Fourier transform: if we write the above quantity $(m^2/2) \int_0^\infty ds(\dots) \int_{-\infty}^\infty dx(\dots)$ as $(m^2/2)\psi_-\psi_+$, we have

$$\psi_- = \left(a + \frac{1}{2} \left\{ \left(\frac{mK_{\mu+1}(m)}{K_\mu(m)} - \mu \right) + \left(\frac{mI_{\lambda-1}(m)}{I_\lambda(m)} - \lambda \right) \right\} \right)^{-1}$$

and

$$\psi_+ = \int_0^\infty dx \left\{ \exp((i\xi + 2)x) \frac{I_\nu(me^{-x})}{I_\nu(m)} + \exp(-(i\xi + 2)x) \frac{K_\nu(me^x)}{K_\nu(m)} \right\}.$$

6. Conclusion

Throughout this paper, we have attempted to show that, with the help of the time-changes discussed in **2.3**, it is possible to 'transfer' certain formulae about the winding number θ_t of planar Brownian motion into formulae about $A_t^{(\nu)}$, especially when time t is being replaced by an independent exponential time. In the course of this process, we were able to give some interpretations, and refinements of several Feynman–Kac formulae which are found in different places in the literature.

Finally, it may be appropriate to mention here two papers which have, at least partly, the same flavor:

(a) Chan (1994) obtains some formulae for the Laplace transforms, at independent exponential times, of occupation times for planar Brownian motion. The excursion theory for planar Brownian motion away from the circle of radius 1 may obviously be derived, thanks to the time-changes in **2.3**, from excursion theory away from 0 for 1-dimensional Brownian motion.

(b) Yor (1984) relates computations about square-root boundaries for Bessel processes to the bird navigation investigations of Kendall (1974), which involve the total winding of certain pole-seeking Brownian motions.

References

Barlow, M.T., Pitman, J.W. and Yor, M. (1989). Une extension multidimensionnelle de la loi de l'arc sinus. *Sém. Probas. XXIII. Lect. Notes in Maths.* no. 1372. Springer, Berlin, 294–314

Biane, Ph. and Yor, M. (1988). Sur la loi des temps locaux browniens pris en un temps exponentiel. *Sém. Probas. XXII. Lect. Notes in Maths.* no. 1321. Springer, Berlin, 454–466

Chan, T. (1994) Occupation times of compact sets by planar Brownian motion. *Annales I.H.P.,* **30**, n° 2, 317–329

De Schepper, A., Goovaerts, M. and Delbaen, F. (1992). The Laplace transform of annuities certain with exponential time distribution. *Insurance: Mathematics and Economics,* **11** (4), 291–304

Geman, H. and Yor, M. (1993). Bessel processes, Asian options and perpetuities *Mathematical Finance,* **3** (4), 349–375. **Paper [5] in this book**

Getoor, R.K. and Sharpe, M.J. (1979). Excursions of Brownian motion and Bessel processes. *Zeitschrift für Wahr.,* **47**, 83–106

Itô, K. and McKean, H.P. (1965). *Diffusion Processes and their Sample Paths.* Springer, Berlin

Kendall, D. (1974). Pole-seeking Brownian motion and bird navigation. *Journal of the Royal Statistical Society,* Series B, **36** (3), 365–417

Lebedev, N. (1972). *Special Functions and their Applications.* Dover Publications, New York

Petit, F. (1992). Quelques extensions de la loi de l'arc sinus. *C.R.A.S. Paris,* t. 315, Série I, 855–858

Pitman, J.W. and Yor, M. (1981). Bessel processes and infinitely divisible laws. In: D. Williams, ed., Stochastic integrals. *Lect. Notes in Maths.* no. 851. Springer, Berlin, 285–370

Spitzer, F. (1958). Some theorems concerning 2-dimensional Brownian motion. *Trans. Amer. Math. Soc.,* **87**, 187–197

Revuz, D. and Yor, M. (1991). *Continuous Martingales and Brownian Motion.* Springer Berlin

Williams, D. (1974). Path decomposition and continuity of local time for one-dimensional diffusions I. *Proc. London Math. Soc. (3),* **28**, 738–768

Yor, M. (1984). On square-root boundaries for Bessel processes, and pole-seeking Brownian motion. In: A. Truman and D. Williams, eds., Stochastic analysis and applications. *Lect. Notes in Maths.* no. 1095. Springer, Berlin

Yor, M. (1992a). On some exponential functionals of Brownian motion. *Advances in App. Proba.,* **24**, 509–531. **Paper [2] in this book**

Yor, M. (1992b). Some aspects of Brownian motion, Part I: Some special functionals. *Lectures in Mathematics, E.T.H. Zürich,* Birkhäuser, Basel

Yor, M. (1992c). Sur les lois des fonctionnelles exponentielles du mouvement brownien, considérées en certains instants aléatoires. *Comptes Rendus Acad. Sci. Paris,* t. 314, Série I, 951–956. **Paper [4] in this book**

Yor, M. (1997). Some aspects of Brownian motion, Part II: Some new martingale problems. *Lectures in Mathematics, E.T.H. Zürich,* Birkhäuser, Basel

Postscript #7

This article presents close relations between the winding number $(\theta_u, u \geq 0)$, more accurately: the continuous determination of the argument of the trajectory of planar Brownian motion $(Z_u, u \geq 0)$ and the exponential functional $(A_t \equiv \int_0^t ds \exp(2B_s), t \geq 0)$ for $(B_s, s \geq 0)$ a real-valued Brownian motion, namely: the inverse process of the clock of $(\theta_u, u \geq 0)$ is precisely $(A_t, t \geq 0)$.

For a recent and historical survey of scales and clocks, see:

H.P. McKean (2001): *"Brownian motion and the general diffusion: scale & clock"*. In: Mathematical Finance - Bachelier Congress 2000. eds: H. Geman, D. Madan, S.R. Pliska and T. Vorst. Springer (2001).

As a consequence, exact and/or asymptotic results concerning θ_u and/or A_t may be transferred from one quantity to the other. Applying a variant of this methodology,

Y. Hu, Z. Shi, M. Yor (1999): *"Rates of convergence of diffusions with drifted Brownian motion potentials"*. Trans. Amer. Math. Soc. **351** n° 10, p. 3915–3934

obtain limit distributions for certain Brownian motions/diffusions in random environments, in agreement with previous results obtained by Kawazu-Tanaka (see, e.g., reference 17. in paper [10] in this book).

On Exponential Functionals of Certain Lévy Processes

Stochastics and Stochastic Rep. **47** (1994), 71–101
(with P. Carmona and F. Petit)

Abstract. In this article we generalize the work of M. Yor concerning the law of $A_T = \int_0^T \exp(\xi_s)ds$ where ξ is a Brownian motion with drift and T an independent exponential time, to the case where ξ belongs to a certain class of Lévy processes. Our method hinges on a bijection, introduced by Lamperti, between exponentials of Lévy processes and semi-stable Markov processes.

1. Introduction

The aim of this work is to study the distribution of the variable $A_t = \int_0^t \exp(\xi_s)\,ds$ for fixed t, where $(\xi_s, s \geq 0)$ denotes a real-valued Lévy process, that is, a process with independent stationary increments.

Such a study, which is important in financial mathematics for calculation of the price of Asian options, was carried out in [20, 21], in the case ξ is a Brownian motion with constant drift, i.e.

$$\xi_t = B_t + vt \quad t \geq 0,$$

where $(B_t, t \geq 0)$ denotes the real-valued Brownian motion.

In this article, in our general study, we rely again on the methods put in place in [20, 21], namely:

- we first seek an explicit expression for the distribution of A_{T_λ} and, more generally, for the pair $(A_{T_\lambda}, \exp(\xi_{T_\lambda}))$, where T_λ is an exponential time with parameter $\lambda > 0$, independent of the process $(\xi_t, t \geq 0)$;
- we reduce this study to that of the determination of the semigroups of certain semi-stable Markov processes.

In a precise manner, the works of Lamperti ([11, 12]) show that there exists a Fellerian process X on $(0, +\infty)$ for which $T_0 = A_\infty$, where:

$$T_0 \stackrel{\text{def}}{=} \inf\{u : X_u = 0\}$$

and which satisfies

$$\exp(\xi_t) = X_{A_t}, \quad \forall t \geq 0. \tag{1.a}$$

Here, X is a semi-stable Markov process on $(0, +\infty)$, which is to say that if $(P_x, x > 0)$ denotes the family of distributions for X, then for all $\lambda > 0$ and $x > 0$, we have:

$$(X_u, u \geq 0, P_x) \stackrel{\text{dist.}}{=} \left(\frac{1}{\lambda} X_{\lambda u}, u \geq 0, P_{\lambda x}\right). \tag{1.b}$$

Conversely, any semi-stable Markov process X with values in $\mathbb{R}_+$ is Fellerian on $(0, +\infty)$ and can be associated with a Lévy process satisfying the relation (1.a) which can also be written as:

$$X_u = \exp \xi_{C_u}, \text{with } C_u = \inf\{s : A_s > u\} = \int_0^u \frac{ds}{X_s}, \quad \forall u < T_0. \tag{1.c}$$

It then appeared natural to study examples of semi-stable Markov processes encountered in various earlier studies ([4, 19, 22]), which apparently had little in common. In fact, these works enable us to exhibit examples of pairs of Markov semi-groups satisfying the intertwining relation:

$$P_t \Lambda = \Lambda Q_t,$$

where Λ denotes a Markov kernel. Such relations arise naturally, either as generalizations of algebraic relations between independent random variables ([22]) or as consequences of conditions under which a function of a Markov process is again a Markov process (Dynkin [7], page 325, Chapter X, Section 6 and Pitman and Rogers [15]).

The class of Lévy processes which appears in these examples is characterized by the Lévy exponents, defined for suitable m, by:

$$E[\exp(m\xi_t)] = \exp(t\psi(m)), \quad t \geq 0,$$

which are of the form

$$\psi(m) = m\frac{am + b}{cm + d}.$$

It is clear that, putting aside the degenerate case $\psi(m) = \alpha m$, which corresponds to $\xi_t = \alpha t$, and the case of Brownian motion with drift, where $\psi(m) = m(am + b)$, our generic Lévy process is of the form

$$\xi_t = \alpha t \pm \text{Pois}(\beta, \gamma)_t, \quad t \geq 0,$$

where $\alpha \in \mathbb{R}$ is the drift term and $\text{Pois}(\beta, \gamma)$ is a compound Poisson process with parameter $\beta > 0$ and jump probability on $(0, +\infty)$, $\nu(dx) = \gamma \exp(-\gamma x)dx$ (Feller [8], pages 288 and 474). More precisely, if $(U_n, n \in \mathbb{N})$ and $(V_n, n \in \mathbb{N})$ are two mutually independent sequences of independent, equidistributed random variables, such that U_n has an exponential distribution with parameter β and V_n has law ν, then the process

$$N_t = \sum_{n=1}^{\infty} 1_{(U_1 + \cdots + U_n \leq t)}; \quad t \geq 0$$

is a Poisson process with parameter β and the process

$$\xi_t = \sum_{n=1}^{N_t} V_n; \quad t \geq 0$$

is a compound Poisson process with parameter β and jump probability ν. Consequently,

- if $\xi_t = \alpha t + \mathrm{Pois}(\beta, \gamma)_t$, then $\psi(m) = m(\alpha + (\beta)/(\gamma - m))$, $m < \gamma$;
- if $\xi_t = \alpha t - \mathrm{Pois}(\beta, \gamma)_t$, then $\psi(m) = m(\alpha - (\beta)/(\gamma + m))$, $m > -\gamma$.

Let us now indicate how our work is organized.

- In Section 2 we use the Girsanov transformation to identify the distribution of A_{T_λ}.
- Section 3 uses intertwinings to calculate quickly the distribution of A_{T_λ}.
- Section 4 presents instances of semi-stable processes of the family then gives a detailed description of the process:

$$X_t^{(\alpha)} = |B_{\tau_t^\mu}|,$$

where $(B_t, t \geq 0)$ is a standard Brownian motion, $\mu = 1/\alpha$, and $(\tau_t^\mu, t \geq 0)$ is the reciprocal of the local time at zero for the semi-martingale $X_t^\mu = |B_t| - \mu l_t$, where $(l_t; t \geq 0)$ denotes the local time at zero of the Brownian motion B (the case $\alpha = 1$ is handled by Watanabe [18]). The properties of other semi-stable Markov processes are deduced from this study.
- In Section 5 we review the explicit distributions for certain random variables associated with these Lévy processes.
- Finally, in the Appendix, we give the proofs of certain results.

2. Girsanov Transformation and Calculation of the Distribution of A_{T_λ}

2.1. General Discussion

We consider a Lévy process $(\xi_t; t \geq 0)$, starting at 0, with $(\mathcal{G}_t)$ its natural filtration and $(\mathbb{P}_a, a \in \mathbb{R})$ the natural family of its distributions, i.e. $\mathbb{P}_a$ is the distribution of $(\xi_t + a; t \geq 0)$. Since $(A_t; t \geq 0)$ is a continuous additive functional of the process ξ, the associated semi-stable Markov process X defined by the relations (1.a) and (1.c) is a Markov process with respect to the filtration $\mathcal{F}_u = \mathcal{G}_{C_u}$ and to the family of probabilities $(P_x = \mathbb{P}_{\log x}; x > 0)$. It follows from the relation (1.a) and from general results on infinitesimal generators, that if $\mathcal{L}$ and L denote the generators of the processes ξ and X, respectively, then they are intertwined by:

$$Lf(x) = (1/x)\mathcal{L}(f \circ \exp)(\log x), \text{ and } \mathcal{L}f(\xi) = e^\xi L(f \circ \log)(e^\xi).$$

We recall that the (Lévy) exponent ψ of ξ is defined by:

$$\mathbb{E}(\exp m\xi_t) = \exp t\psi(m).$$

For the family of processes studied, these quantities are equal to:

- if $\xi_t = \alpha t + \mathrm{Pois}(\beta, \gamma)_t$, then $\psi(m) = m(\alpha + (\beta)/(\gamma - m))$, $m < \gamma$ and:

$$Lf(x) = \alpha f'(x) + (\beta\gamma/x) \int_1^\infty (f(xy) - f(x))dy/y^{\gamma+1}$$

$$\mathcal{L}f(\xi) = \alpha f'(\xi) + \beta\gamma \int_0^\infty (f(\xi + \eta) - f(\xi))e^{-\gamma\eta}d\eta$$

- if $\xi_t = \alpha t - \mathrm{Pois}(\beta, \gamma)_t$, then $\psi(m) = m(\alpha - (\beta)/(\gamma + m))$, $m > -\gamma$ and:

$$Lf(x) = \alpha f'(x) + (\beta\gamma/x) \int_0^1 (f(xy) - f(x))y^{\gamma-1}dy$$

$$\mathcal{L}f(\xi) = \alpha f'(\xi) + \beta\gamma \int_0^\infty (f(\xi - \eta) - f(\xi))e^{-\gamma\eta}d\eta.$$

When $\psi(m) < \infty$, the process $(\exp(m\xi_t - t\psi(m)); t \geq 0)$ is a $\mathcal{G}_t$ martingale and we can consider the Girsanov transform $\mathbb{P}_a^{(m)}$ of the probability $\mathbb{P}_a$ defined by

$$\mathbb{P}_a^{(m)} = \exp(m(\xi_t - a) - t\psi(m)) \cdot \mathbb{P}_a, \quad \text{on } \mathcal{G}_t.$$

The relations between the processes X and ξ enable us to state the following proposition.

Proposition 2.1. *Under the probability* $\mathbb{P}_a^{(m)}$, *the process* ξ *is a Lévy process with exponent given by*

$$\psi^{(m)}(p) = \psi(m + p) - \psi(m).$$

The associated semi-stable Markov process $X^{(m)}$ *is, in fact, the process* X *considered under the new probability:*

$$P_x^{(m)} = (X_u/x)^m \exp(-\psi(m)C_u) \cdot P_x, \quad \text{on } \mathcal{F}_u.$$

We call this type of transformation a Girsanov power transformation. The following proposition links A_{T_λ} to the family of random variables (H_p) defined, when they exist, by:

$$P(H_p > t) = E_1 \left(\frac{1}{X_t^p} \right).$$

Of course, $(H_p^{(m)})$ denotes the family associated with the Girsanov transformed process $X^{(m)}$.

Proposition 2.2. *Suppose $\lambda > 0$ is such that there exists m satisfying $\psi(m) = \lambda$. Then, $H_m^{(m)}$ exists and*

$$A_{T_\lambda} \overset{\text{dist.}}{=} H_m^{(m)}.$$

Proof. For all $u \geq 0$, with a time change, followed by a Girsanov power transformation, we have:

$$\mathbb{P}_0(A_{T_\lambda} > u) = \lambda \int_0^\infty \exp(-\lambda t)\mathbb{P}_0(A_t > u)dt$$

$$= E_1\left(\int_{C_u}^\infty \lambda \exp(-\lambda t)dt\right) = E_1(\exp -\lambda C_u)$$

$$= E_1(\exp -\psi(m)C_u) = E_1^{(m)}\left(\frac{1}{(X_u)^m}\right).$$

Explicit knowledge of the semigroup $P_u^{(m)}(x,dy) = P_x^{(m)}(X_u \in dy)$ of the process $X^{(m)}$ will then give us the distribution for the pair $(\exp(\xi_{T_\lambda}), A_{T_\lambda})$. We denote $p_u^{(m)}(dy) = P_u^{(m)}(1,dy)$. Because of the semi-stability, we have $P_u^{(m)}(x,dy) = p_{u/x}^{(m)}(dy/x)$. It follows from Proposition 2.1 that if $\psi(m) < \infty$ then

$$x^m E_1(\exp(-\psi(m)C_u)|X_u = x)p_u(dx) = p_u^{(m)}(dx).$$

Consequently, $p_u^{(m)}(dx)$ has a density, which we shall denote by $\delta_u^{(m)}(x)$, with respect to $p_u(dx)$ and:

$$E_1(\exp(-\psi(m)C_u)|X_u = x) = \frac{\delta_u^{(m)}(x)}{x^m}.$$

Thus, if $\lambda = \psi(m) \in (0,\infty)$, $(\delta_u^{(m)}(x)/x^m)$ is the Laplace transform in λ of a probability measure on $\mathbb{R}_+$:

$$\phi_{u,x}(dy) = P(C_u \in dy|X_u = x).$$

In order to eliminate some possible ambiguity and to simplify the calculations, in what follows, we shall choose for m the unique $x > 0$ such that $\lambda = \psi(x)$.

Proposition 2.3. *If $\lambda = \psi(m) \in (0,\infty)$, then*

$$\mathbb{P}_0(\exp(\xi_{T_\lambda}) \in dx, A_{T_\lambda} \in du) = \lambda \, du \, \frac{p_u^{(m)}(dx)}{x^{1+m}}.$$

Moreover, $\phi_{u,x}(dy) = \phi_{u,x}(y)dy$, and:

$$\mathbb{P}_0(\exp(\xi_t) \in dx, A_t \in du) = \frac{du}{x}\phi_{u,x}(t)p_u(dx).$$

Proof. Let $f : \mathbb{R}_+ \times \mathbb{R}_+ \to \mathbb{R}_+$ be a bounded continuous function. By time change and Girsanov power transformation, we have

$$
\mathbb{E}_0(f(\exp(\xi_{T_\lambda}), A_{T_\lambda})) = \lambda E_1 \left(\int_0^\infty \exp(-\lambda C_u) f(X_u, u) dC_u \right)
$$

$$
= \lambda E_1 \left(\int_0^\infty \exp(-\psi(m) C_u) f(X_u, u) \frac{du}{X_u} \right)
$$

$$
= \lambda E_1^{(m)} \left(\int_0^\infty f(X_u, u) \frac{du}{X_u^{1+m}} \right)
$$

$$
= \lambda \int_0^\infty \int f(x, u) \frac{p_u^{(m)}(dx)}{x^{1+m}} du
$$

and the first part of the proposition is proved. We rewrite this equality by dividing the two extreme terms of the above chain of equalities by λ, to obtain:

$$
\int_0^\infty e^{-\lambda t} \mathbb{E}_0(f(\exp(\xi_t), A_t)) dt = \int_0^\infty \int_0^\infty f(x, u) \frac{p_u^{(m)}(dx)}{x^{1+m}} du
$$

$$
= \int_0^\infty \int_0^\infty f(x, u) p_u(dx) \frac{du}{x} \frac{\delta_u^{(m)}(x)}{x^m}
$$

$$
= \int_0^\infty e^{-\lambda t} \int \int \phi_{u,x}(dt) p_u(dx) \frac{du}{x} f(x, u),
$$

which gives the second part of the proposition, by virtue of the injectivity of the Laplace transform.

2.2. The Example of Brownian Motion with Drift

We recall two results of [20] (Proposition 6.1 and Theorem 6.2), whose proofs used arguments analogous to those of Subsection **2.1**.

Proposition 2.4. *Given $\mu \in \mathbb{R}$ and $n \in \mathbb{N}$, there exists a polynomial P_n such that:*

$$
\mathbb{E}_0 \left(\exp(\mu \xi_t) \left(\int_0^t \exp(\xi_s) ds \right)^n \right) = E \left[\exp(\mu \xi_t) P_n(\exp(\xi_t)) \right],
$$

with

$$
P_n(z) = n! \sum_0^n c_j z^j, \text{ and } c_j = \prod_{k \neq j, 0 \leq k \leq n} (\psi(\mu + j) - \psi(\mu + k))^{-1}.
$$

For any real $k \geq 0$, we have:

$$
\mathbb{E}_x(((A_{T_\lambda} - k)^+)^n) = n \int_k^\infty (u - k)^{n-1} E_x[\exp(-\lambda C_u)] du.
$$

Remark. The moments of A_t were obtained prior to [20] by D. Dufresne ([5, 6]) with methods entirely different to those of [20].

Let us now introduce the squared Bessel process of dimension 2α, denoted by $(X_\alpha(t))_{t\geq 0}$, whose infinitesimal generator is:

$$L_\alpha f(x) = 2x f''(x) + 2\alpha f'(x).$$

We denote by Q_t^α the semigroup of the process $(X_\alpha(t))_{t\geq 0}$. It is given by the two formulae

$$Q_t^\alpha f(0) = \int_0^{+\infty} \frac{y^{\alpha-1}}{\Gamma(\alpha)(2t)^\alpha} \exp\left(-\frac{y}{2t}\right) f(y) dy$$

$$Q_t^\alpha f(x) = \int_0^{+\infty} \frac{1}{2t} \left(\frac{y}{x}\right)^{(\alpha-1)/2} \exp\left(-\frac{x+y}{2t}\right) I_{\alpha-1}\left(\sqrt{xy}/t\right) f(y) dy, \quad x > 0.$$

It is easy to see that this is a Fellerian process on $\mathbb{R}_+$. We note that the family of processes $(X_\alpha(t))_{t\geq 0}$ is, in fact, the only family of continuous semi-stable Markov processes.

To pursue our illustration of Subsection **2.1** in the case of Brownian motion with drift, we require some notation.

Notation. 1. When k is a natural number and λ an arbitrary real number, by convention we set $\Gamma(\lambda+k)/\Gamma(\lambda) = 1$ if $k = 0$, and $\Gamma(\lambda+k)/\Gamma(\lambda) = \lambda(\lambda+1)\cdots(\lambda+k-1)$ otherwise.

2. Z_a denotes a positive random variable having a gamma distribution with parameter $a > 0$, that is:

$$\mathbb{P}[Z_a \in dx] = \frac{x^{a-1}}{\Gamma(a)} e^{-x} 1_{\{x>0\}} dx.$$

3. $Z_{a,b}$ denotes a positive random variable having a beta distribution with parameters $a > 0$ and $b > 0$, that is:

$$\mathbb{P}[Z_{a,b} \in dx] = \frac{x^{a-1}(1-x)^{b-1}}{B(a;b)} 1_{\{0<x<1\}} dx.$$

4. Finally, if A, B and C are random variables, the notation

$$A \overset{\text{dist.}}{=} BC$$

signifies that A is identical in distribution to the product of two independent random variables with the same distributions as B and C. We recall, for example, the result:

$$Z_a \overset{\text{dist.}}{=} Z_{a,b} Z_{a+b}.$$

Proposition 2.5. *If X_α is the squared Bessel process of dimension $2\alpha > 0$, we have:*

$$p_u(dx) = (2u)^{-1} x^{(\alpha-1)/2} I_{\alpha-1}(\sqrt{x}/u) \exp\left(-\frac{1+x}{2u}\right) 1_{x>0} dx$$

$$p_u^{(m)}(dx) = (2u)^{-1} x^{(\alpha+2m-1)/2} I_{\alpha+2m-1}(\sqrt{x}/u) \exp\left(-\frac{1+x}{2u}\right) 1_{x>0} dx.$$

In what follows, m denotes the unique strictly positive real number satisfying $\lambda = \psi(m) = 2m(m+\alpha-1)$, namely: $m = m(\lambda) = \frac{1}{2}(1-\alpha+\sqrt{2\lambda+(\alpha-1)^2})$. Then, if I_ν denotes the modified Bessel function of index $\nu = \sqrt{2\lambda+(\alpha-1)^2}$:

$$\mathbb{P}(\exp(\xi_{T_\lambda}) \in dx, A_{T_\lambda} \in dt)$$
$$= \frac{\lambda}{2t} x^{(\alpha-3)/2} I_\nu(\sqrt{x}/t) \exp\left(-\frac{1+x}{2t}\right) 1_{\{x>0; t>0\}} dx dt$$

$$\mathbb{P}(\exp(\xi_{T_\lambda}) \in dx) = \frac{\lambda}{2} \frac{x^{\alpha+m-2}}{2m+\alpha-1} (1 \vee x)^{1-2m\alpha} 1_{x>0} dx$$

$$\xi_t = 2(B_t + (\alpha-1)t).$$

If $m > n > 0$:

$$\mathbb{E}[((A_{T_\lambda} - k)^+)^n]$$
$$= 2^{-m} \frac{\Gamma(1+n)\Gamma(\alpha+m)\Gamma(m-n)k^{n-m}}{\Gamma(\alpha+2m)\Gamma(m)} {}_1F_1\left(m-n; \alpha+2m; -\frac{1}{2k}\right)$$
$$\tag{2.a}$$

$$A_{T_\lambda} \overset{\text{dist.}}{=} Z_{1,\alpha+m-1}/(2Z_m). \tag{2.b}$$

Letting λ tend to 0, in the case $\alpha < 1$, we obtain:

$$A_\infty \overset{\text{dist.}}{=} 1/(2Z_{1-\alpha}).$$

Denoting:

$$P_n^\mu(z) = (-2)^{-n} \sum_0^n (-z)^j C_n^j (2\mu+\alpha-1+2j) \frac{\Gamma(2\mu+\alpha-1+j)}{\Gamma(2\mu+\alpha+j+n)},$$

where $C_n^j \equiv \binom{n}{j}$

$$\mathbb{E}\left[(\exp \mu \xi_t)\left(\int_0^t \exp \xi_s ds\right)^n\right] = \mathbb{E}[(\exp \mu \xi_t)P_n^\mu(\exp \xi_t)].$$

$$E_1[\exp(-\lambda C_u)|X_u = x] = I_\nu(\sqrt{x}/u)/I_{\alpha-1}(\sqrt{x}/u)$$

$$E_1[\exp(-\lambda C_u)] = \frac{\Gamma(\alpha+m)}{\Gamma(\alpha+2m)}(2u)^{-m}\,_1F_1\left(m;\alpha+2m;-\frac{1}{2u}\right).$$

Remarks. 1. Letting k tend to 0 in (2.a) we find:

$$\mathbb{E}[(A_{T_\lambda})^n] = 2^{-n}\frac{\Gamma(1+n)\Gamma(\alpha+m)\Gamma(m-n)}{\Gamma(\alpha+m+n)\Gamma(m)}$$

which is consistent with (2.b).

2. Inverting the expression for $\mathbb{E}[(A_{T_\lambda})^n]$ in λ, we find

$$\mathbb{E}[(A_t)^n] = \frac{n!}{\Gamma(2n)}\int\int_{\substack{0\leq x\leq t\\ s\geq 0}} (1-e^{-s})^{2n-1}\frac{\,_2F_1(n,2n+\alpha;2n;1-e^{-s})}{\exp(\frac{1+\alpha}{2}s+\frac{s^2}{8x}+\frac{(\alpha-1)^2}{2}x)}\cdots$$

$$\cdots\frac{sdxds}{x^{3/2}2^{n+1}(2\pi)^{1/2}}.$$

3. In the case $\alpha = 1$, $\mu = 0$, the expression for P_n^μ can be simplified. We find:

$$P_n^0(z) = (-2)^{-n}\frac{2}{n!}\left(\,_2F_1(-n,1;n+1;z)-1\right).$$

3. Intertwinings of Semi-stable Markov Processes

One of the simplest ways of determining the families (H_p) and, as a repercussion, the distribution of A_{T_λ}, involves the use of the notion of intertwining between two Markov processes.

Definition. Given two semigroups $(P_t; t \geq 0)$ and $(Q_t; t \geq 0)$, these semigroups are said to be *intertwined by the Markov kernel* Λ if

$$P_t\Lambda = \Lambda Q_t, \quad t \geq 0.$$

We denote by $C_0(\mathbb{R}_+)$ the space of continuous functions on $\mathbb{R}_+$ which are zero at infinity. Λ is said to be the *multiplicative kernel associated with the positive random variable M* if

$$\Lambda f(x) = E(f(Mx)) \quad \forall f \in C_0(\mathbb{R}_+).$$

Suppose we have two semi-stable Markov processes X and Y, with semi-groups $(P_t; t \geq 0)$ and $(Q_t; t \geq 0)$, respectively. Consider the families (H_p) and (K_p) of random variables defined, when they exist, by:

$$P(H_p > t) = E_1(X_t^{-p}), \quad P(K_p > t) = E_1(Y_t^{-p}).$$

Proposition 3.1. *Suppose that $E[M^{-p}] < +\infty$. If H_p or K_p exists and if*

$$P_t \Lambda = \Lambda Q_t, \quad t \geq 0,$$

then both H_p and K_p exist and

$$P(H_p > t) = \frac{1}{E[M^{-p}]} E[M^{-p} 1_{(MK_p > t)}].$$

Proof. This is a simple consequence of the intertwining relation applied to the function $f_{-p}(x) = x^{-p}$.

Remark. For certain classes of variables M, the previous relation can be simplified:

- if $M = Z_a$, then for all $p < a$, we have $H_p \overset{\text{dist.}}{=} Z_{a-p} K_p$;
- if $M = Z_{a,b}$, then for all $p < a$, we have: $H_p \overset{\text{dist.}}{=} Z_{a-p,b} K_p$.

It suffices to note that, for example, in the first case:

$$E[M^{-p} 1_{(MK_p > t)}] = E\left[\int_{t/K_p}^\infty \frac{x^{a-p-1}}{\Gamma(a)} e^{-x} dx \right] = \frac{\Gamma(a-p)}{\Gamma(a)} P(Z_{a-p} K_p > t).$$

Another important consequence of the intertwining relation and a sufficient condition for intertwining are given by the following proposition.

Proposition 3.2. *Let Λ be the multiplicative kernel associated with the positive random variable M. Suppose we have two semi-stable Fellerian processes, X and Y on $\mathbb{R}_+$ with semigroups $(P_t; t \geq 0)$ and $(Q_t; t \geq 0)$, respectively, with respective infinitesimal generators L^X and L^Y, whose domains we shall denote by $D(L^X)$ and $D(L^Y)$. Suppose that the distribution of X, starting at 0, is determinate, that is, if f and g are functions of $C_0(\mathbb{R}_+)$ satisfying:*

$$\forall t \geq 0, \quad P_t f(0) = P_t g(0) \text{ then: } f = g.$$

Then the following assertions are equivalent:

1. $P_t \Lambda = \Lambda Q_t$;
2. *when X and Y start from 0, we have:*

$$Y_1 \overset{\text{dist.}}{=} M X_1;$$

3. $\forall f \in D(L^Y)$, $\Lambda f \in D(L^X)$, *and* $L^X \Lambda f = \Lambda L^Y f$.

Proof. (1) $\Rightarrow$ (2). For all $f \in C_0(\mathbb{R}_+)$, we have:

$$E_0(f(M X_1)) = P_1 \Lambda f(0) = \Lambda Q_1 f(0) = Q_1 f(0) = E_0(f(Y_1)).$$

(2) $\Rightarrow$ (3). First, we note that because of the semi-stability, for all $f \in C_0(\mathbb{R}_+)$ we have:

$$\begin{aligned} P_t \Lambda f(0) &= E_0(f(M X_t)) = E_0(f(M t X_1)) = E_0(f(t Y_1)) = E_0(f(Y_t)) \\ &= Q_t f(0) = \Lambda Q_t f(0). \end{aligned}$$

Let $f \in D(L^Y)$; for all $s \geq 0$, $t > 0$:

$$\begin{aligned} P_s \left(\frac{P_t \Lambda f - \Lambda f}{t} \right)(0) &= \frac{P_{t+s}\Lambda f(0) - P_s \Lambda f(0)}{t} = \frac{\Lambda Q_{t+s} f(0) - \Lambda Q_s f(0)}{t} \\ &= \Lambda Q_s \left(\frac{Q_t f - f}{t} \right)(0). \end{aligned}$$

Letting t tend to 0, we obtain, for the term on the right-hand side of this equality:

$$\Lambda Q_s L^Y f(0) = P_s \Lambda L^X f(0).$$

Let us fix $a \geq 0$. We have just shown that the function $g(t) = P_{a+t}\Lambda f(0)$ is continuous and right differentiable on $\mathbb{R}_+$ with right derivative $g'_d(t) = P_{a+t}\Lambda L^Y f(0)$. Since this right derivative is continuous, it follows ([14], Corollary 1.2, page 43) that g is of class C^1 on $\mathbb{R}_+$. Consequently for all $t \geq 0$:

$$\begin{aligned} P_a (P_t \Lambda f - \Lambda f)(0) = g(t) - g(0) &= \int_0^t g'(s)ds = \int_0^t ds P_{a+s}\Lambda L^Y f(0) \\ &= P_a \left(\int_0^t P_s \Lambda L^Y f ds \right)(0). \end{aligned}$$

This equality holds for all $a \geq 0$; since, by hypothesis, the distribution of $(X_t; t \geq 0)$, starting at 0, is determinate, we obtain

$$P_t \Lambda f - \Lambda f = \int_0^t P_s \Lambda L^Y f ds, \quad \forall t \geq 0.$$

It follows by continuity that:

$$\lim_{t \to 0} \frac{P_t \Lambda f - \Lambda f}{t} = \lim_{t \to 0} \frac{1}{t} \int_0^t P_s \Lambda L^Y f \, ds = \Lambda L^Y f,$$

which implies that $\Lambda f \in D(L^X)$ and $L^X \Lambda f = \Lambda L^Y f$.

(3) $\Rightarrow$ (1). Let $(R_\lambda, \lambda > 0)$ and $(W_\lambda, \lambda > 0)$ be the respective families of resolvents for X and Y. Let $g \in C_0(\mathbb{R}_+)$ and $\lambda > 0$. Then

$$f = W_\lambda g \text{ is in } D(L^Y) \text{ and } g = (\lambda I - L^Y) f.$$

By hypothesis, $\Lambda f \in D(L^X)$ and $L^X \Lambda f = \Lambda L^Y f$. Consequently:

$$\Lambda g = \lambda \Lambda f - \Lambda L^Y f = \lambda \Lambda f - L^X \Lambda f = (\lambda I - L^X) \Lambda f.$$

By definition of the resolvent, this means that

$$R_\lambda \Lambda g = \Lambda f = \Lambda W_\lambda g,$$

which can be rewritten as:

$$\int_0^\infty e^{-\lambda t} P_t \Lambda g \, dt = \int_0^\infty e^{-\lambda t} \Lambda Q_t g \, dt, \quad \forall \lambda > 0.$$

Since the functions $t \to P_t \Lambda g$ and $t \to \Lambda Q_t g$ are continuous, we conclude, by virtue of the injectivity of the Laplace transform, that:

$$P_t \Lambda g = \Lambda Q_t g, \quad \forall t \geq 0, \quad \forall g \in C_0(\mathbb{R}_+).$$

Proposition 3.3, below, will enable us to determine the distribution of X, starting at 0, from the exponent ψ. Proposition 3.4 will provide us with a practical means of verifying the intertwining relations.

Proposition 3.3. *Let X be a semi-stable Markov process on $\mathbb{R}_+$, associated with the Lévy process ξ of exponent ψ. We suppose that ψ is defined on a neighbourhood of a real p such that $\psi(p) < 0$. Then:*

$$E_0(X_1^p) = \begin{cases} (\psi(p)/p) E_0(X_1^{p-1}), & \text{if } p < 0 \\ 0, & \text{if } p > 0. \end{cases}$$

Consequently, if for some interval of values of p, we have $p > 0$ and $\psi(p) < 0$, then the process X, starting at 0, remains at 0; if, in addition, X is strongly Markov, then X is absorbed at 0.

Remark. If $p > 0$ and $\psi(p) < 0$, then $\lim_{t \to \infty} \xi_t = -\infty$, which explains why X is absorbed at 0.

Proof. Let us first show that

$$\psi(p) < 0 \Rightarrow \int_0^\infty E_1(X_t^{p-1})dt = -1/\psi(p)$$

$$\psi(p) > 0 \Rightarrow \int_0^\infty E_1(X_t^{p-1})dt = +\infty.$$

In effect, by time change, we have:

$$\int_0^\infty E_1(X_t^{p-1})dt = E_1\left(\int_0^\infty X_t^p dC_t\right) = \mathbb{E}_0\left(\int_0^\infty \exp(p\xi_u)du\right)$$

$$= \int_0^\infty \exp(\psi(p)u)du.$$

Let now p be such that $\psi(p) < 0$. The process:

$$t \mapsto \exp(p\xi_t) - 1 - \psi(p)\int_0^t \exp(p\xi_s)ds$$

is a $(\mathcal{G}_t)$ local martingale under $\mathbb{P}_0$. Now, $(C_t; t \geq 0)$ is an increasing family of $(\mathcal{G}_t)$ stopping times, hence

$$M_t = X_t^p - \psi(p)\int_0^t ds X_s^{p-1}$$

is a $(\mathcal{F}_t)$ local martingale. Let (T_n) be a sequence of stopping times reducing M; we have:

$$E_1(X_{t \wedge T_n}^p) = 1 + \psi(p)E_1\left(\int_0^{t \wedge T_n} X_u^{p-1}du\right)$$

$$= (-\psi(p))E_1\left(\int_{t \wedge T_n}^\infty du X_u^{p-1}\right)$$

$$\leq (-\psi(p))E_1\left(\int_0^\infty du X_u^{p-1}\right)$$

$$\leq 1.$$

Consequently, by Fatou's Lemma:

$$\sup_{t \geq 0} E_1(X_t^p) \leq 1.$$

Now, ψ is continuous wherever it is defined (in fact ψ is convex); thus, there exists $\varepsilon > 0$ such that $\psi(p(1 + \varepsilon)) < 0$, and

$$\sup_{t \geq 0} E_1(X_t^{p(1+\varepsilon)}) \leq 1.$$

It follows from this that $(X_t^p; t \geq 0)$ is a uniformly integrable supermartingale, that $t \to E_1(X_t^p)$ is right continuous, that H_{-p} exists, and,

$$E_1(X_t^p) = (-\psi(p)) \int_t^\infty du E_1(X_u^{p-1}).$$

Denoting $f_p(x) = x^p$, by semi-stability, using the previous result, we have

$$P_t f_p(x) = E_1((xX_{t/x})^p) = f_p(x) + \psi(p) \int_0^t dv P_v f_{p-1}(x).$$

Consequently, since X is semi-stable, we have $P_t f(0) = E_0(f(tX_1))$, and

$$P_{s+t} f_p(0) = P_s P_t f_p(0) = P_s f_p(0) + \psi(p) \int_0^t dv\, P_{s+v} f_{p-1}(0).$$

Since $P_r f_p(0) = r^p E_0(X_1^p)$, it follows that:

$$(s+t)^p E_0(X_1^p) - s^p E_0(X_1^p) = (\psi(p)/p)((s+t)^p - s^p) E_0(X_1^{p-1}).$$

If $p > 0$, since $\psi(p) < 0$, the terms on the two sides of this equality have opposite signs; thus, we must have $E_0(X_1^p) = E_0(X_1^{p-1}) = 0$. On the other hand, if $p < 0$, things can be simplified to obtain $E_0(X_1^p) = (\psi(p)/p) E_0(X_1^{p-1})$.

Remark. One immediate consequence of this proposition is that H_p exists if $\psi(-p) < 0$. This result agrees with Proposition 2.2, for if (K_p) denotes the family associated with the process with exponent $\phi(p) = \psi(m - p) - \psi(-p)$, then $H_p = K_p^{(p)}$ exists if $\phi(p) = -\psi(-p) > 0$.

A priori, a semi-stable Markov process X with semigroup (P_t) is Fellerian on $(0, \infty)$ in the sense of Lamperti ([12], Lemma 2.1); that is, for all $f \in C_0(\mathbb{R}_+)$ the function $P_t f(x)$ is continuous on $(0, \infty)$ and tends to 0 at infinity. We shall say that the process X is Fellerian on $\mathbb{R}_+$ if its semigroup can be extended to a Fellerian semigroup on $\mathbb{R}_+$; for this, it is sufficient that for all $f \in C_0(\mathbb{R}_+)$ the function $P_t f(x)$ has a limit when $x \to 0$. Λ again denotes the multiplicative kernel associated with the random variable M.

Proposition 3.4. *Let X and Y be two semi-stable processes on $(0, \infty)$, with semigroups $(P_t; t \geq 0)$ and $(Q_t; t \geq 0)$, respectively, and exponents ψ and $\hat{\psi}$, respectively. Let us suppose there exist two real numbers $0 \leq a < b$ such that*

$$\forall p \in (-b, -a), \quad \psi(p) < 0, \ \hat{\psi}(p) < 0, \quad E(M^p) = (\psi(p)/\hat{\psi}(p)) E(M^{p-1}).$$

Then:

$$\forall f \in C_0(\mathbb{R}_+), \ \forall x > 0, \quad P_t \Lambda f(x) = \Lambda Q_t f(x). \tag{3.a}$$

Moreover, X and Y are Fellerian on $\mathbb{R}_+$, and we have

$$\forall f \in C_0(\mathbb{R}_+), \quad P_t \Lambda f = \Lambda Q_t f \tag{3.b}$$

if one of the following three conditions is satisfied:

1. X and Y are Fellerian on $\mathbb{R}_+$;
2. X is Fellerian on $\mathbb{R}_+$, and $M = Z_a$ or $M = 1/Z_a$ $(a > 0)$;
3. Y is Fellerian on $\mathbb{R}_+$, and Λ is an injective operator on $C_0(\mathbb{R}_+)$.

Proof. Let $p \in (-b, -a)$. We establish, as in the proof of the previous proposition, that:

$$\forall x > 0, \quad P_t f_p(x) = E_x(X_t^p) = (-\psi(-p)) \int_t^\infty dv \, P_v f_{p-1}(x) < \infty.$$

Consequently, since $\Lambda f_p = E(M^p) f_p$,

$$\forall x > 0, \quad P_t \Lambda f_p(x) = E(M^p)(-\psi(-p)) \int_t^\infty dv \, P_v f_{p-1}(x).$$

On the other hand, by Fubini's Theorem:

$$\Lambda Q_t f_p(x) = (-\hat{\psi}(p)) \int_t^\infty \Lambda Q_v f_{p-1}(x) dv$$

$$= E(M^{p-1})(-\hat{\psi}(p)) \int_t^\infty Q_v f_{p-1}(x) dv.$$

Consequently:

$$P_t \Lambda f_p(x) = \Lambda Q_t f_p(x), \quad \forall x > 0, \ t \geq 0, \ p \in (-b, -a).$$

The expression (3.a) follows easily from the identity:

$$f_{-p}(y) = \frac{1}{y^p} = \int_0^\infty \frac{z^{p-1}}{\Gamma(p)} e^{-zy} dz,$$

by an argument involving the injectivity of the Laplace transform.

To prove (3.b) it is sufficient to prove that the two processes are Fellerian on $\mathbb{R}_+$ and to pass to the limit as x tends to 0.

1. The two processes are Fellerian on $\mathbb{R}_+$ by hypothesis.
2. Let us suppose that X is Fellerian on $\mathbb{R}_+$. For all $f \in C_0(\mathbb{R}_+)$ we have:

$$\Lambda Q_t f(x) = P_t \Lambda f(x) \to P_t \Lambda f(0), \quad \text{when } x \text{ tends to } 0.$$

Lemma 3.5 tells us that $Q_t f(x)$ converges when x tends to 0, and thus allows to conclude that (Q_t) is a Fellerian semigroup on $\mathbb{R}_+$.

3. Let us now suppose that Y is Fellerian on $\mathbb{R}_+$, and that Λ is injective. For all $f \in C_0(\mathbb{R}_+)$ we have:

$$P_t \Lambda f(x) = \Lambda Q_t f(x) \to \Lambda Q_t f(0) = Q_t f(0), \quad \text{when } x \text{ tends to } 0.$$

It is therefore sufficient to show that the image of Λ is dense in $C_0(\mathbb{R}_+)$. Let μ be a Radon measure on $\mathbb{R}_+$, orthogonal to the image of Λ. For all $f \in C_0(\mathbb{R}_+)$, and all $\lambda \geq 0$, the function $f_\lambda(x) = f(\lambda x)$ is in $C_0(\mathbb{R}_+)$, hence:

$$0 = \int \mu(dx) \Lambda f_\lambda(x) = \int \mu(dx) \int_0^\infty P(M \in dy) f(\lambda x y) = \Lambda g(\lambda),$$

where the function $g(y) = \int f(xy)\mu(dx)$ is in $C_0(\mathbb{R}_+)$. By the injectivity of Λ, we have $g = 0$, and since f is arbitrary, the measure μ is zero.

Remark. A sufficient condition for the kernel Λ to be injective is ([2], Theorem 4.8.4):

$$E[M^{iu}] \neq 0, \quad \forall u \in \mathbb{R}.$$

Lemma 3.5. If $M = Z_a$ or $M = 1/Z_a$ and if g is a continuous bounded function on $(0, \infty)$ such that $\Lambda g(x) \to k$ when $x \to 0$, then $g(x) \to k$ when $x \to 0$.

Proof. Let us suppose, for example, that $M = Z_a$. We have:

$$\Lambda g(x) = E(g(x Z_a)) = \frac{1}{\Gamma(a)x^a} \int_0^\infty g(y) y^{a-1} e^{-y/x} dy \to k, \quad \text{when } x \to 0.$$

We set $U(y) = g(y)y^{a-1}$ and $V(\lambda) = \lambda \int_0^\infty e^{-\lambda t} U(t)dt$; it follows that

$$V(\lambda) \sim k\,\Gamma(a)\lambda^{1-a}, \quad \text{when } \lambda \to \infty.$$

Consequently ([2], Theorem 1.7.1), $U(y) \sim ky^{a-1}$, when $y \to 0$, that is $g(y) \to k$ when $y \to 0$.

4. Description of the Semi-stable Markov Processes being Studied

4.1. Realization of some Semi-stable Markov Processes

In a natural way, we seek to construct examples of semi-stable Markov processes with the help of Brownian motion in $\mathbb{R}^n$, which leads us to consider the square of the norm of the Brownian motion, that is the Bessel square of dimension n. The Bessel square of (arbitrary) dimension 2α, which is denoted by $(X_\alpha(t))_{t \geq 0}$ (see Proposition 2.5) and which generalizes the above, is linked to the Itô measure of Brownian excursions via the Lévy–Khintchine representation of its infinitely divisible distribution. If $(B_t)_{t\geq0}$ is a standard Brownian motion, and if l_t denotes its local time at zero at the instant t, we find again, for example, the Bessel squares in the description of the family of local times of the μ-process $X_t^\mu \equiv |B_t| - \mu l_t$ ($\mu > 0$), taken at τ_t^μ, the right-continuous reciprocal of the local time at zero of X^μ ([1, 4, 13]). With the help of the μ-process, we then construct another semi-stable Markov process denoted by $X^{(\alpha)}(t) \equiv |B_{\tau_t^\mu}| = \mu l_{\tau_t^\mu}$ with $\mu = 1/\alpha$ ([4]). In Section 4.2. we shall study in detail this process, whose semigroup (also Fellerian on $\mathbb{R}_+$, like all the examples studied here) and its generator, are given by the following proposition, the proof of which is given in the Appendix.

Proposition 4.1. 1) *The semigroup of the process* $(X^{(\alpha)}(t); t \geq 0)$ *is given by:*

$$T_t^\alpha f(x) = f(x)e^{-t/2x}$$
$$+ \frac{t\alpha}{2}x^{\alpha-1}e^{-t/2x}\int_x^{+\infty} {}_1F_1\left(1+\alpha; 2; \frac{t(y-x)}{2xy}\right)f(y)y^{-\alpha-1}dy.$$

2) *The generator of* $(X^{(\alpha)}(t); t \geq 0)$ *is:*

$$L^{(\alpha)}f(x) = \frac{1}{2x}E[f(x/Z_{\alpha,1}) - f(x)] = \frac{\alpha}{2x}\int_1^{+\infty}\frac{f(xy) - f(x)}{y^{\alpha+1}}dy$$

for any function f *of class* $\mathcal{C}^\infty$ *with compact support in* $\mathbb{R}_+$.

Let us now consider the process

$$X^{1,\beta}(t) \equiv t + \sup_{s\leq\tau_{t\beta}}(|B_s| - (1/\beta)l_s),$$

where $(\tau_s)_{s\geq0}$ is the right-continuous inverse of the local time l. This gives us a new example ([4]). If we apply to this process the Girsanov transformations considered in Proposition 2.1 (the Girsanov power transformations), we obtain the processes $X^{\alpha,\beta}(t)$ whose associated semigroups are given explicitly by:

$$S_t^{\alpha,\beta}f(x) = \left(\frac{x}{x+t}\right)^\beta f(x+t) + \frac{\alpha\beta t}{(x+t)^{1-\alpha}}\int_{x+t}^{+\infty}\frac{(y-t)^{\beta-1}}{y^{\alpha+\beta}}\cdots$$
$$\cdots {}_2F_1\left(1-\beta, 1-\alpha; 2; \frac{t(y-x-t)}{(x+t)(y-t)}\right)f(y)dy$$

and whose infinitesimal generators are given by

$$L^{\alpha,\beta}f(x) = f'(x) + \frac{\alpha\beta}{x}\int_1^{+\infty}\frac{f(xy) - f(x)}{y^{\alpha+1}}dy$$

for any function f of class $\mathcal{C}^\infty$ with compact support in $\mathbb{R}_+$.

Finally, the Markov extension ([19, 22]) of the algebraic relation:

$$Z_\alpha \stackrel{\text{dist.}}{=} Z_{\alpha,\beta} Z_{\alpha+\beta}$$

leads to the appearance when $\alpha + \beta \geq 1$, of the semigroups $\Pi_t^{\alpha,\beta}$ and $\hat{\Pi}_t^{\alpha,\beta}$, which are dual to each other for the measure $\mu(dx) = 1_{x>0} x^{\alpha-1} dx$, that is, they satisfy: for all Borel functions $f, g : \mathbb{R}_+ \to \mathbb{R}_+$,

$$\int \Pi_t^{\alpha,\beta} f(x) g(x) \mu(dx) = \int f(x) \hat{\Pi}_t^{\alpha,\beta} g(x) \mu(dx).$$

The semigroup $\Pi_t^{\alpha,\beta}$, whose associated process is denoted by $X_{\alpha,\beta}(t)$, satisfies:

$$\Pi_t^{\alpha,\beta} f(x) = \left(\frac{x}{x+t} \right)^\beta f(x+t)$$

$$+ \left(\frac{x}{x+t} \right)^\beta \int_0^1 y^{\alpha+\beta-2} f((t+x)y) G_{\alpha,\beta} \left(\frac{t(1-y)}{xy} \right) dy$$

where the function $G_{\alpha,\beta}$ is defined by

$$G_{\alpha,\beta}(u) = \begin{cases} (u^{\beta-1}/B(\alpha;\beta))\, {}_2F_1\left(-\beta, 1-\beta; \alpha; \frac{1}{u}\right) & \text{if } u > 1 \\ \beta(\alpha+\beta-1)\, {}_2F_1(2-\alpha-\beta, 1-\beta; 2; u) & \text{if } u < 1. \end{cases}$$

Its infinitesimal generator is given by

$$L_{\alpha,\beta} f(x) = f'(x) + \frac{\beta(\alpha+\beta-1)}{x} \int_0^1 y^{\alpha+\beta-2} (f(xy) - f(x)) dy$$

for any function f of class $\mathcal{C}^\infty$ with compact support in $\mathbb{R}_+$. The semigroup $\hat{\Pi}_t^{\alpha,\beta}$, whose associated process is denoted by $\hat{X}_{\alpha,\beta}(t)$ satisfies:

$$\hat{\Pi}_t^{\alpha,\beta} f(x) = 1_{x>t} \left(\frac{x-t}{x} \right)^{\alpha+\beta-1} f(x-t)$$

$$+ \int_0^{t \wedge x} \left(\frac{x-u}{x} \right)^{\alpha+\beta-1} \frac{u^{\beta-2}}{t^{\beta-1}} f\left(t\frac{x-u}{u} \right) G_{\alpha,\beta} \left(\frac{t-u}{x-u} \right) du.$$

Its infinitesimal generator is given by:

$$\hat{L}_{\alpha,\beta} f(x) = -f'(x) + \beta \frac{\alpha+\beta-1}{x} \int_1^{+\infty} \frac{f(xy) - f(x)}{y^{1+\beta}} dy$$

for any function f of class $\mathcal{C}^\infty$ with compact support in $\mathbb{R}_+$.

4.2.   Description of $X^{(\alpha)}(t)$

In Section 4.3. we shall see just how useful it is to have a good knowledge not only of the Brownian motion with drift, but also of one particular process in the family of semi-stable Markov processes which we are studying. That is why we describe in detail here the process $X^{(\alpha)}(t)$ which we met in our previous studies and whose definition we now recall:

$$X^{(\alpha)}(t) = |B_{\tau_t^\mu}| = \mu l_{\tau_t^\mu}, \quad (\mu = 1/\alpha),$$

where τ_t^μ is the right-continuous inverse of the local time at zero of $X_t^\mu \equiv |B_t| - \mu l_t$.

We know ([16], Chapter II, Exercise 19, page 80, for example) that if S_t denotes the supremum of a standard Brownian motion on $[0, t]$, then the process $(S_{\tau_t})_{t\geq 0}$ studied by Watanabe in [18] is a Markov process. We call this the Watanabe process. The identity in distribution due to Paul Lévy $(S_t - B_t; S_t)_{t\geq 0} \overset{\text{dist.}}{=} (|B_t|; l_t)_{t\geq 0}$ shows that the Watanabe process can be represented in the form $X^{(\alpha)}(t)$ with $\alpha = 1$.

Proposition 4.2. *We set* $\alpha = 1/\mu$. *The process* $(X^{(\alpha)}(t) \equiv |B(\tau_t^\mu)| = \mu l(\tau_t^\mu); t \geq 0)$ *is a homogeneous Markov process. We call this a generalized Watanabe process.*

Proof. We set $Z_t = (|B_t|; l_t)$. Then Z, a two-dimensional process, is a strong Markov process ([16], Chapter VI, Exercise (2.18), page 227; caution, the μ-process is not Markov for $\mu \neq 1$). Thus, for any process obtained by changing the time of Z via the inverse of an additive continuous functional of Z is again strongly Markov ([3], Chapter V, Proposition 2.11, page 212). Taking for the additive functional the local time at zero of the μ-process, we obtain that $(Z(\tau_t^\mu); t \geq 0)$ is a Markov process whose components are linked via the relation $|B_{\tau_t^\mu}| = \mu l_{\tau_t^\mu}$. The stated result follows from this.

Let us now recall the expressions for the semigroup T^α and for the infinitesimal generator $L^{(\alpha)}$ of the process $(X^{(\alpha)}(t); t \geq 0)$ (see Proposition 4.1):

$$T_t^\alpha f(x) = f(x)e^{-t/2x}$$

$$+ \frac{t\alpha}{2} x^{\alpha-1} e^{-t/2x} \int_x^{+\infty} {}_1F_1\left(1 + \alpha; 2; \frac{t(y-x)}{2xy}\right) f(y) y^{-\alpha-1} dy$$

$$L^{(\alpha)} f(x) = \frac{\alpha}{2x} \int_1^{+\infty} \frac{f(xy) - f(x)}{y^{\alpha+1}} dy$$

(for any function f of class $\mathcal{C}^\infty$ with compact support in $\mathbb{R}_+$).

Here are some remarks arising from these expressions.

1. It is not necessary to calculate the semigroup T_t^α explicitly in order to know the expression of its generator. In fact, noting that $\tau_0^{\mu,x}$ is just $T^{\mu,x}$, the first passage time through x for the μ-process, we obtain:

$$\frac{1}{t}(T_t^\alpha f(x) - f(x)) = \frac{e^{-t/2x}}{2x} \frac{1}{t} \int_0^t e^{s/2x} E[f(x + \mu l_{\tau_s^{\mu,x}}) - f(x)]ds,$$

which implies

$$L^{(\alpha)}f(x) = \frac{1}{2x}E_0[f(|B_{T^{\mu,x}}|) - f(x)] = \frac{1}{2x}E[f(x/Z_{\alpha,1}) - f(x)].$$

The distribution of $|B_{T^{\mu,x}}|$ can be found, for example, using the solution of the Skorokhod problem given in [9] (pages 210 to 225); then we have:

$$|B_{T^{\mu,x}}| \stackrel{\text{dist.}}{=} \frac{x}{Z_{\alpha,1}}.$$

2. In the case $\alpha < 1$, the expression for the semigroup can be simplified and becomes:

$$T_t^\alpha f(x) = E\left\{ f\left(\frac{(x \vee t/2Z_1) - xZ_{1-\alpha,\alpha}}{1 - Z_{1-\alpha,\alpha}} \right) \right\}$$

where the beta and gamma variables involved are independent. When α tends to 1, we rediscover the semigroup of the Watanabe process:

$$T_t^1 f(x) = E[f(x \vee t/2Z_1)].$$

3. The semigroup T_t^α can also be written in the form:

$$T_t^\alpha f(x) = f(x)e^{-t/2x} + \int_0^{t/2x} e^{-z}Q_{t-2xz}^2[f(x + Y_x/2Z_{1/\mu})]dz$$

where Y is a Bessel square of dimension two (starting at $t - 2xz$).

4. The semi-stable Markov process $(X^{(\alpha)}(t))_{t\geq 0}$ is a Fellerian process on $\mathbb{R}_+$ (the semi-stability property is evident by virtue of the scaling properties of the Brownian motion). In fact, the only difference lies in the continuity at zero, which can be proved 'by hand' from the explicit expression for the semigroup.

5. If B is a Brownian motion starting at $x > 0$, and if $\tau_t^{\mu,x}$ is the inverse of the local time at zero of the process $(|B_s| - \mu l_s(B))_{s\geq 0}$, then $(|B(\tau_t^{\mu,x})|)_{t\geq 0}$ has the same distribution as $(X^{(\alpha)}(t))_{t\geq 0}$. In fact, let us denote by $l_t^a(H) = \int_0^t \delta(H_u - a)du$ the local time of a process H at level a and at time t, and by P_t the semigroup of the Markov process $(Z(\tau_t^\mu); t \geq 0)$ (see the proof of Proposition 4.1). Since the components of $(Z(\tau_t^\mu); t \geq 0)$ are related, if $g(a; b) = f(a)$, we have:

$$T_t^\alpha f(x) = P_t g(x; \alpha x).$$

Now,

$$P_t g(x; y) = E_{(x,y)}[g((X; Y)_{\inf\{s; \int_0^s \delta(X_u - \mu Y_u)du > t\}})]$$

where X and Y are the components of Z.

$$P_t g(x; y) = E_x[g((|B|; y + l^0(B))_{\inf\{s; \int_0^s \delta(|B_u| - \mu y - \mu l_u^0(B))du > t\}})]$$
$$= E_x[g(|B(\lambda_t)|; y + l_{\lambda_t}^0(B))],$$

where λ_t is the inverse of the local time at the level μy of the μ-process constructed from a Brownian motion starting at x. The result follows from this.

6. The explicit expression of the generator $L^{(\alpha)}$ allows us to calculate the exponent ψ_α of the associated Lévy process, by virtue of the relation $L^{(\alpha)} f_m = \psi_\alpha(m) f_m$, which we shall prove in the Appendix. Thus:

$$\psi_\alpha(m) = \frac{m}{2(\alpha - m)} \quad \text{for } m < \alpha.$$

This exponent is that of the compound Poisson process with parameter $1/2$, for which the jump probability is the exponential distribution with parameter α. In other terms, the Lévy process associated with the semi-stable Markov process is:

$$\xi_t = \text{Pois}\left(\frac{1}{2}; \alpha\right)_t.$$

7. Finally, on reading the semi-group, one sees that the processes obtained from $(X^{(\alpha)}(t))_{t\geq 0}$ using a Girsanov transformation of the powertransformation type (Proposition 2.1) have the same distributions as $(X^{(\alpha-m)}(\alpha t/(\alpha - m)))_{t\geq 0}$. For $m < \alpha$, the density $p_t^{(m)}(dx)$ is thus obtained by changing α to $(\alpha - m)$ and t to $(\alpha t)/(\alpha - m)$ in the expression for $p_t(dx)$. Thus, we remain in the same family of processes when we carry out a power transformation; this is a very important property for the rest of our study.

It then remains to apply the theoretical results of Propositions 2.3 and 2.4, to obtain the following explicit results.

Proposition 4.3. *The explicit expressions for the quantities associated with the process* $(X^{(\alpha)}(t))_{t\geq 0}$ *are:*

$$p_u(dx) = e^{-u/2}\varepsilon_1(dx) + \frac{u\alpha}{2}e^{-u/2} \, {}_1F_1\left(1 + \alpha; 2; u\frac{x-1}{2x}\right) x^{-(1+\alpha)} 1_{x>1} dx.$$

Setting $m = m(\lambda) = (2\lambda\alpha)/(1 + 2\lambda)$, *we have:*

$$\mathbb{P}(\exp(\xi_{T_\lambda}) \in dx, A_{T_\lambda} \in dt)$$

$$= \lambda 1_{t>0} \exp\left(-t\left(\lambda + \frac{1}{2}\right)\right) dt \left(\varepsilon_1(dx) + \frac{t\alpha}{2}x^{-(\alpha+2)}\right.$$

$$\left.\times {}_1F_1\left(1 + \frac{\alpha}{1+2\lambda}; 2; \left(\frac{1}{2} + \lambda\right)t\frac{x-1}{x}\right)1_{x>1}dx\right)$$

$$\mathbb{P}(A_{T_\lambda} \in dt) = \lambda \exp\left(-t\left(\lambda + \frac{1}{2}\right)\right) {}_1F_1\left(\frac{\alpha}{1+2\lambda}; 1+\alpha; \left(\frac{1}{2}+\lambda\right)t\right)1_{t>0}dt$$

$$\mathbb{P}(\exp(\xi_{T_\lambda}) \in dx) = \frac{2\lambda}{1+2\lambda}\varepsilon_1(dx) + 2\lambda\alpha(1+2\lambda)^{-2}x^{-(1+m)}1_{x>1}dx$$

$$\mathbb{E}[\exp - \lambda\xi_t] = \exp - \frac{\lambda t}{2(\alpha+\lambda)}$$

that is:

$$\mathbb{P}(\xi_t \in dx) = e^{-t/2}\left\{\varepsilon_0(dx) + \frac{t\alpha}{2}e^{-\alpha x}{}_0F_1\left(2; \frac{t\alpha x}{2}\right)1_{x>0}dx\right\}$$

$$= e^{-t/2}\left\{\varepsilon_0(dx) + \frac{1}{2}e^{-\alpha x}\sqrt{\frac{t\alpha}{x}}I_1(\sqrt{2t\alpha x})1_{x>0}dx\right\}$$

$$\mathbb{E}_1[\exp(-\lambda C_u)|X_u = x]$$

$$= e^{-\lambda u}\frac{{}_1F_1(1 + \alpha/(1+2\lambda); 2; (\lambda+1/2)u(1-1/x))}{{}_1F_1(1+\alpha; 2; (1-1/x)u/2)} \qquad x \geq 1$$

$$\mathbb{E}_1[\exp(-\lambda C_u)] = {}_1F_1\left(\frac{2\lambda\alpha}{1+2\lambda}; \alpha; -\frac{1+2\lambda}{2}u\right).$$

If $\mu + n < \alpha$, *denoting*

$$P_n^\mu(z) = -\left(\frac{2}{\alpha}\right)^n \sum_0^n (-z)^j C_n^j(\alpha - \mu - j)^{n-1}\frac{\Gamma(n+1+\mu-\alpha)}{\Gamma(\mu-\alpha)},$$

where $C_n^j \equiv \binom{n}{j}$, *we have:*

$$\mathbb{E}\left[(\exp\mu\xi_t)\left(\int_0^t \exp\xi_s ds\right)^n\right] = \mathbb{E}[(\exp\mu\xi_t)P_n^\mu(\exp\xi_t)]$$

$$A_{T_\lambda} \stackrel{\text{dist.}}{=} \frac{2Z_1}{(1+2\lambda)Z_{2\lambda\alpha/(1+2\lambda),\alpha/(1+2\lambda)}}.$$

If $n < \alpha$, for all $\lambda > \psi(n)$, we have:

$$\mathbb{E}[((A_{T_\lambda} - k)^+)^n] = \frac{n!}{\lambda} \left(\lambda + \frac{1}{2}\right)^{1-n} \frac{\Gamma(-n + 2\lambda\alpha/(1 + 2\lambda))}{\Gamma(\alpha - n)} \cdots$$
$$\cdots {}_1F_1\left(\frac{2\lambda\alpha}{1 + 2\lambda} - n; \alpha - n; -\frac{1 + 2\lambda}{2}k\right).$$

4.3. Study of Lévy Processes of the Form $\xi_t = \alpha t \pm \mathrm{Pois}(\beta, \gamma)_t$

We refer the reader to the start of Section 2 for a description of the general form of the infinitesimal generator of ξ. To begin with, we recall in Table A, the processes of the family to which the examples of semi-stable Markov processes studied in detail in Section 5 correspond.

   This section justifies the preceding theoretical study: a knowledge of the associated family (H_p) for two semi-stable Markov processes, will enable us to calculate *algebraically* the distributions of the A_{T_λ} for all other processes of the family by virtue of the stability of this family under the Girsanov power transformation. We denote the Lévy process under study by ξ and its associated semi-stable Markov process by X. We continue to take $p > 0$. The results are summarized in Table B.

Remark. One might be surprised at the constraints on the parameters α, β, γ in lines 3, 4 and 5 of Table B. It is easy to see that the cases not studied correspond to processes for which there exists an interval of values of $p > 0$ for which $\psi(p) < 0$. It follows from Proposition 3.4 that the only possible extension to $\mathbb{R}_+$ of the semi-stable Markov process X is then the trivial extension with absorption at 0!.

Setting up Table B. Let us first consider the process $X^{(\alpha)}$ which corresponds to the Lévy process $\xi_t = \mathrm{Pois}(1/2, \alpha)_t$. The explicit form for the semigroup enables us to affirm that $X^{(\alpha)}$ is a Fellerian process on $\mathbb{R}_+$ such that:

$$(X_1, P_0) \overset{\mathrm{dist.}}{=} \frac{1}{2Z_\alpha}, \quad H_p \overset{\mathrm{dist.}}{=} \frac{2Z_1}{Z_{p,\alpha}}, \quad \forall p > 0.$$

The linear transformation of time $\xi_t \to \xi_{at}$ corresponds to the transformation $X_t \to X_{at}$ and therefore, the semi-stable Markov process X associated with the Lévy process $\xi = \mathrm{Pois}(\beta, \gamma)$ is a Fellerian process on $\mathbb{R}_+$, for which the distribution of (X_1, P_0) and the family (H_p) are given in the second line of Table B.

   Suppose we have $\alpha_1 > 0$ and $0 < \alpha_2 < \beta/\gamma$. We denote X_1, respectively X_2, the semi-stable Markov process associated with

$$\xi_1(t) = \alpha_1 t + \mathrm{Pois}(\beta, \gamma)_t, \text{respectively } \xi_2(t) = -\alpha_2 t + \mathrm{Pois}(\beta, \gamma)_t;$$

Table B

ξ_t	X_1 under P_0	H_p	A_{T_λ}	Validity
$2((\alpha-1)t+B_t)$	$2Z_\alpha$	$Z_{1,\alpha-p-1}/(2Z_p)$	$Z_{1,\alpha+m-1}/(2Z_m)$	$\lambda=2m(m+\alpha-1)$ $m>\sup(0,1-\alpha)$
$\text{Pois}(\beta,\gamma)_t$	β/Z_γ	$Z_1/(\beta Z_{p,\gamma})$	$Z_{1,\gamma-m}/(\beta Z_{m,\gamma-m})$	$\lambda=m\beta/(\gamma-m)$ $0<m<\gamma$
$\alpha t+\text{Pois}(\beta,\gamma)_t$ $\alpha>0$	$\alpha/Z_{\gamma,\beta/\alpha}$	$Z_1/(\alpha Z_{p,\alpha}Z_{p+\gamma+\beta/\alpha})$	$Z_1/(\alpha Z_{m,\alpha}Z_{\gamma+(\gamma\beta/\alpha(\gamma-m))})$	$\lambda=m(\alpha+(\beta/(\gamma-m)))$ $0<m<\gamma$
$\alpha t-\text{Pois}(\beta,\gamma)_t$ $\alpha>\beta/\gamma$	$\alpha Z_{\gamma+1-\beta/\alpha,\beta/\alpha}/Z_\gamma$	$(Z_{\gamma+1}/Z_p)Z_{1,\gamma-p-\beta/\alpha}$ $0<p<\inf(\gamma+1,\gamma-\beta/\alpha)$	$(Z_{\gamma+1}/Z_m)Z_{1,\gamma-(\gamma\beta/\alpha(\gamma+m))}$	$\lambda=m(\alpha-(\beta/(\gamma+m)))$ $m>0$
$-\alpha t+\text{Pois}(\beta,\gamma)_t$ $0<\alpha<\beta/\gamma$	$\alpha Z_{1-\gamma+\beta/\alpha}/Z_\gamma$	$Z_{1,-p-\gamma+\beta/\alpha}/(\alpha Z_{p,\gamma})$ $0<p<\beta/\gamma$	$Z_{1,-\gamma+(\beta/\alpha(\gamma-m))}/(\alpha Z_{m,\gamma-m})$	$\lambda=m(-\alpha+(\beta/(\gamma-m)))$ $0<m<\gamma$

Table A

P_t	Exponent ψ	Lévy process ξ
Q_t^α	$2m(m+\alpha-1)$	$2((\alpha-1)t+B_t)$
T_t^α	$m/[2(\alpha-m)]$	$\text{Pois}(1/2,\alpha)_t$
$S_t^{\alpha,\beta}$	$m(\alpha+\beta-m)/(\alpha-m)$	$t+\text{Pois}(\beta,\alpha)_t$
$\Pi_t^{\alpha,\beta}$	$m(\alpha+m-1)/(\alpha+\beta+m-1)$	$t-\text{Pois}(\beta,\alpha+\beta-1)_t$
$\hat\Pi_t^{\alpha,\beta}$	$m(\alpha+m-1)/(\beta-m)$	$-t+\text{Pois}(\alpha+\beta-1,\beta)_t$

Table C

$Y_1 \stackrel{\text{dist.}}{=} MX_1$	$P_t\Lambda=\Lambda Q_t$
$2Z_\alpha \stackrel{\text{dist.}}{=} Z_{\alpha,\beta}(2Z_{\alpha+\beta})$	$Q_t^{\alpha+\beta}\Lambda_{\alpha,\beta}=\Lambda_{\alpha,\beta}Q_t^\alpha$
$2Z_\alpha \stackrel{\text{dist.}}{=} (2Z_{\alpha+\beta})Z_{\alpha,\beta}$	$\Pi_t^{\alpha,\beta}\Lambda_{\alpha+\beta}=\Lambda_{\alpha+\beta}Q_t^\alpha$
$Z_\alpha/Z_\beta \stackrel{\text{dist.}}{=} (2Z_\alpha)(1/2Z_\beta)$	$T_t^\beta\Lambda_\alpha=\Lambda_\alpha\hat\Pi_t^{\alpha,\beta}$
$Z_\alpha/Z_\beta \stackrel{\text{dist.}}{=} (1/2Z_\beta)(2Z_\alpha)$	$Q_t^\alpha\tilde\Lambda_\beta=\tilde\Lambda_\beta\hat\Pi_t^{\alpha,\beta}$
$(1/2Z_\alpha) \stackrel{\text{dist.}}{=} (1/2Z_{\alpha+\beta})(1/Z_{\alpha,\beta})$	$S_t^{\alpha,\beta}\tilde\Lambda_{\alpha+\beta}=\tilde\Lambda_{\alpha+\beta}T_t^\alpha$
$(1/2Z_\alpha) \stackrel{\text{dist.}}{=} (1/Z_{\alpha,\beta})(1/2Z_{\alpha+\beta})$	$T_t^{\alpha+\beta}\tilde\Lambda_{\alpha,\beta}=\tilde\Lambda_{\alpha,\beta}T_t^\alpha$

thus, for all $m < \gamma$ we have:

$$\psi(m) = m\beta/(\gamma - m), \quad \psi_1(m) = m(\alpha_1 + \beta/(\gamma - m)),$$
$$\psi_2(m) = m(-\alpha_2 + \beta/(\gamma - m)).$$

Finally, we consider the random variables $M_1 = \beta/(\alpha_1 Z_{\gamma+\beta/\alpha_1})$ and $M_2 = (\alpha_2/\beta)Z_{(\beta/\alpha_2)-\gamma+1}$. If Λ_1 and Λ_2 are the associated multiplicative kernels, we shall prove the following intertwining relations:

$$P_t^1 \Lambda_1 = \Lambda_1 P_t, \quad P_t \Lambda_2 = \Lambda_2 P_t^2.$$

These two relations are simple consequences of Proposition 3.4, because X is a Fellerian process and for all $p < 0$ we have: $\psi(p) < 0$, $\psi_1(p) < 0$ and

$$\frac{E(M_1^p)}{E(M_1^{p-1})} = \frac{\psi(p)}{\psi_1(p)} = \frac{\beta}{\alpha_1(\gamma + \beta/\alpha_1 - p)}.$$

On the other hand, for all $p \in (\gamma - \beta/\alpha_2, 0)$, we have: $\psi_2(p) < 0$, $\psi(p) < 0$ and

$$\frac{E(M_2^p)}{E(M_2^{p-1})} = \frac{\psi_2(p)}{\psi(p)} = (\alpha_2/\beta)(p - \gamma + \beta/\alpha_2).$$

Following Propositions 3.1 and 3.2, the above two intertwining relations enable us to obtain the distributions of (X_1, P_0) and (H_p) of lines 3 and 5 of Table B. To obtain the distributions for the variables A_{T_λ}, we note that the Girsanov power transformation changes the process $\xi_t = \alpha t + \text{Pois}(\beta, \gamma)_t$ to $\xi_t^{(m)} = \alpha t + \text{Pois}(\beta\gamma/(\gamma-m), \gamma-m)_t$ and we apply the relation $A_{T_\lambda} \overset{\text{dist.}}{=} H_m^{(m)}$ for all $\lambda = \psi(m) > 0$.

The derivation of the fourth line of Table B is just as direct. Given $\beta > 0$ and $\alpha > 1$, $X^{\alpha,\beta}$ is the unique Fellerian process on $\mathbb{R}_+$ whose semigroup $(\Pi_t^{\alpha,\beta}; t \geq 0)$ satisfies the intertwining relation

$$\Pi_t^{\alpha,\beta} \Lambda_{\alpha+\beta} = \Lambda_{\alpha+\beta} Q_t^\alpha,$$

where $(Q_t^\alpha; t \geq 0)$ is the semigroup of the Bessel square of dimension 2α, and $\Lambda_{\alpha+\beta}$ is the multiplicative kernel associated with the variable $2Z_{\alpha+\beta}$. (See Section 4.1.). It follows from Propositions 3.1 and 3.2 and the knowledge of the Bessel squares that:

$$(X^{\alpha,\beta}, P_0) \overset{\text{dist.}}{=} Z_{\alpha,\beta}, \quad H_p \overset{\text{dist.}}{=} \frac{Z_{\alpha+\beta-p} Z_{1,\alpha-p-1}}{Z_p}, \quad \forall p \in (0, \alpha - 1).$$

The equalities in distribution relating to (X_1, P_0) and (H_p) corresponding to the fourth line of the table are now deduced from the above by linear transformation in time. We conclude by remarking that the Girsanov power transformation changes the process $\xi_t = \alpha t - \mathrm{Pois}(\beta, \gamma)_t$, with $\alpha > \beta/\gamma$, into the process $\xi_t^{(m)} = \alpha t - \mathrm{Pois}(\beta\gamma/(\gamma + m), \gamma + m)_t$ for all $m > -\gamma$.

Table C summarizes the intertwining relations between the processes of the family: some of these come from the previous proof; the remainder are an immediate consequence of Proposition 3.2. We introduce the following notation:

- X_t and Y_t are the Markov processes associated with the semigroups P_t and Q_t, respectively;
- $\Lambda_\alpha, \tilde{\Lambda}_\alpha, \Lambda_{\alpha,\beta}, \tilde{\Lambda}_{\alpha,\beta}$ are the kernels for multiplications by the variables $2Z_\alpha$, $1/2Z_\alpha, Z_{\alpha,\beta}, 1/Z_{\alpha,\beta}$, respectively.

5. Complementary Explicit Results

In Section 4.3 we saw how to deduce a large number of results concerning a process of the family of semi-stable Markov processes which we have just studied, from those obtained for just two of the processes, the Brownian motion with (sufficiently large) drift and the generalized Watanabe process, for example. In this section, we collect together explicit results which concern the other examples cited. We thus obtain the following propositions.

Proposition 5.1. *In the case of the semigroup $S_t^{\alpha,\beta}$, we have:*

$$p_u(dx) = (1 + u)^{-\beta}\varepsilon_{1+u}(dx) + \alpha\beta u(1 + u)^{\alpha-1}(x - u)^{\beta-1}x^{-(\alpha+\beta)}$$

$$\times \,_2F_1\left(1 - \beta, 1 - \alpha; 2; \frac{u}{1+u}\frac{x - 1 - u}{x - u}\right)1_{x>1+u}dx$$

$$p_u^{(m)}(dx) = (1 + u)^{-\beta\alpha/(\alpha-m)}\varepsilon_{1+u}(dx)$$

$$+ \alpha\beta u\frac{(1 + u)^{\alpha-m-1}}{x - u}\left(\frac{x - u}{x}\right)^{\beta\alpha/(\alpha-m)}x^{m-\alpha}$$

$$\times \,_2F_1\left(1-\beta\alpha/(\alpha-m), 1-\alpha+m; 2; \frac{u}{1+u}\frac{x-1-u}{x-u}\right)1_{x>1+u}dx$$

In what follows, m denotes the unique strictly positive real number satisfying $\lambda = \psi(m) = m(1 + (\beta)/(\alpha - m))$, namely: $m = m(\lambda) = \frac{1}{2}(\alpha + \beta + \lambda - \sqrt{(\alpha + \beta + \lambda)^2 - 4\alpha\lambda})$. Then, if $x \geq 1 + t$, we have:

$$E_1[\exp(-\lambda C_u)|X_u = x]$$

$$= \left(\frac{x - u}{x}\right)^{\lambda-m}(1 + u)^{-m}\frac{\,_2F_1\left(1 - \beta - \lambda + m, 1 - \alpha + m; 2; \frac{u}{1+u}\frac{x-1-u}{x-u}\right)}{\,_2F_1\left(1 - \beta, 1 - \alpha; 2; \frac{u}{1+u}\frac{x-1-u}{x-u}\right)}$$

$$E_1[\exp(-\lambda C_u)]$$

$$= (1+u)^{-\lambda-\beta} + \beta\alpha u(1+u)^{-\lambda-\beta-1} \int_0^1 (1-y)^{\alpha-1} \ldots$$

$$\left(1 - \frac{u}{1+u}y\right)^{m-\alpha-\beta-\lambda} {}_2F_1\left(1-\beta-\lambda+m, 1-\alpha+m; 2; \frac{u}{1+u}y\right) dy$$

$$\mathbb{P}(\exp(\xi_{T_\lambda}) \in dx, A_{T_\lambda} \in dt)$$

$$= 1_{t>0}dt\lambda\left\{(1+t)^{-1-\beta-\lambda}\varepsilon_{1+t}(dx) + \alpha\beta t\frac{(1+t)^{\alpha-m-1}}{x-t}\left(\frac{x-t}{x}\right)^{\beta+\lambda-m} x^{-1-\alpha}\right.$$

$$\left.\ldots {}_2F_1\left(1-\beta-\lambda+m, 1-\alpha+m; 2; \frac{t}{1+t}\frac{x-1-t}{x-t}\right) 1_{x>1+t}dx\right\}.$$

For $n < \alpha - \mu$, we denote:

$$P_n^\mu(z) = (-1)^n \sum_0^n (-z)^j C_n^j \Gamma(a_j)(j+\mu-\alpha)^{-n} \frac{j+a_j}{\Gamma(n+1+a_j)},$$

where $C_n^j \equiv \binom{n}{j}$, and where $a_j \equiv \mu - \alpha + (\alpha\beta/(j+\mu-\alpha))$.

$$\mathbb{E}\left[(\exp\mu\xi_t)\left(\int_0^t \exp\xi_s ds\right)^n\right] = \mathbb{E}[(\exp\mu\xi_t)P_n^\mu(\exp\xi_t)].$$

Proposition 5.2. *In the case of the semigroup $\Pi_t^{\alpha,\beta}$, we have:*

$$p_u(dx) = (1+u)^{-\beta}\varepsilon_{1+u}(dx)$$

$$+ (1+u)^{1-2\beta-\alpha}1_{0<x<1+u}x^{\alpha+\beta-1}G_{\alpha,\beta}\left(u\frac{1+u-x}{x}\right)dx$$

$$p_u^{(m)}(dx) = (1+u)^{-\beta(\alpha+\beta-1)/(\alpha+\beta+m-1)}\varepsilon_{1+u}(dx)$$

$$+ (1+u)^{\alpha+m+2\beta-\beta m/(\alpha+\beta+m-1)}1_{0<x<1+u}x^{\alpha+\beta+m-1} \ldots$$

$$\ldots G_{\alpha+m+\beta m/(\alpha+\beta+m-1),\beta(\alpha+\beta-1)/(\alpha+\beta+m-1)}\left(u\frac{1+u-x}{x}\right)dx$$

In what follows, m denotes the unique strictly positive real number satisfying $\lambda = \psi(m) = m(\alpha+m-1)/(\alpha+\beta+m-1)$, namely: $m = m(\lambda) = \frac{1}{2}(\lambda+1-\alpha+\sqrt{(\lambda+1-\alpha)^2+4\lambda(\alpha+\beta-1)}$. Then

$$A_{T_\lambda} \stackrel{\text{dist.}}{=} Z_{\alpha+\beta}Z_{1,\alpha+m-\lambda-1}/Z_m$$

Letting λ tend to 0, we obtain:

$$A_\infty \overset{\text{dist.}}{=} \begin{cases} Z_{\alpha+\beta}/Z_{1-\alpha} & \text{if } \alpha+\beta \geq 1 \text{ and } \alpha < 1 \\ +\infty & \text{if } \alpha \geq 1. \end{cases}$$

In the case where $\alpha + \beta + \mu > 1$, if we denote

$$P_n^\mu(z) = (-1)^n \sum_0^n (-z)^j C_n^j \frac{\Gamma(\alpha+\beta+\mu+n)}{\Gamma(\alpha+\beta+\mu-1)} \cdots$$

$$\cdots \left(1 - \frac{\beta(\alpha+\beta-1)}{(\alpha+\beta+\mu-1+j)^2}\right) \frac{\Gamma\left(\alpha+\beta+\mu-1-\frac{\beta(\alpha+\beta-1)}{(\alpha+\beta+\mu-1+j)}\right)}{\Gamma\left(n+\alpha+\beta+\mu-\frac{\beta(\alpha+\beta-1)}{\alpha+\beta+\mu-1+j}\right)}$$

then we have:

$$\mathbb{E}\left[(\exp \mu\xi_t)\left(\int_0^t \exp \xi_s ds\right)^n\right] = \mathbb{E}[(\exp \mu\xi_t)P_n^\mu(\exp \xi_t)].$$

Proposition 5.3. *In the case of the semigroup $\hat{\Pi}_t^{\alpha,\beta}$, we have:*

$$p_u(dx) = 1_{0<u<1}(1-u)^{\alpha+\beta-1}\varepsilon_{1-u}(dx)$$
$$+ 1_{x>(1-u)}x^{\alpha+\beta-1}(x+u)^{\alpha+2\beta-1}G_{\alpha,\beta}\left(u\frac{x-1+u}{x}\right)dx$$

$$p_u^m(dx) = 1_{0<u<1}(1-u)^{\beta(\alpha+\beta-1)/(\beta-m)}\varepsilon_{1-u}(dx)$$
$$+ 1_{x>(1-u)}x^{\beta(\alpha+\beta-1)/(\beta-m)}(x+u)^{1+m-\beta+2\beta(\alpha+\beta-1)/(\beta-m)}$$
$$\cdots G_{1+m-\beta+\beta(\alpha+\beta-1)/(\beta-m),\beta(\alpha+\beta-1)/(\beta-m)}\left(u\frac{x-1+u}{x}\right)dx.$$

In what follows, m denotes the unique strictly positive real number satisfying $\lambda = \psi(m) = m(\alpha + m - 1)/(\beta - m)$, namely $m = m(\lambda) = \frac{1}{2}(1 - \alpha - \lambda + \sqrt{(\lambda+\alpha-1)^2 + 4\lambda\beta})$. Then,

$$A_{T_\lambda} \overset{\text{dist.}}{=} Z_{1,\beta(\alpha+m-1)/(\beta-m)}/Z_{m,\beta-m}.$$

Letting λ tend to 0 in the case where $\alpha < 1$, we obtain:

$$A_\infty \overset{\text{dist.}}{=} 1/Z_{1-\alpha,\alpha+\beta-1}.$$

For any integer $n < \beta - \mu$, if we denote

$$P_n^\mu(z) = \sum_0^n (-z)^j C_n^j \frac{\Gamma(\mu - \beta + n + 1)}{\Gamma(\mu - \beta)} \left(1 - \frac{\beta(\alpha + \beta - 1)}{(j + \mu - \beta)^2}\right) \cdots$$

$$\cdots \frac{\Gamma\left(\mu - \beta - \frac{\beta(\alpha + \beta - 1)}{j + \mu - \beta}\right)}{\Gamma\left(n + 1 + \mu - \beta - \frac{\beta(\alpha + \beta - 1)}{j + \mu - \beta}\right)}$$

then, we have:

$$\mathbb{E}\left[(\exp \mu \xi_t)\left(\int_0^t \exp \xi_s ds\right)^n\right] = \mathbb{E}[(\exp \mu \xi_t)P_n^\mu(\exp \xi_t)].$$

Appendix. Proofs of Certain Results

Lemma. *If the two Markov processes X and Y with extended generators L^X and L^Y, respectively, are linked by the relation*

$$h(X_t) = Y\left(\int_0^t k(X_s)ds\right)$$

where h and $k > 0$ are two positive, bijective functions, then:

$$k(h^{-1}(y))L^Y f(y) = L^X(f \circ h)(h^{-1}(y)).$$

Proof. ([7]). We set $\alpha_t \equiv \int_0^t k(X_s)ds$, and $a_t \equiv \inf\{s; \alpha_s > t\}$. Let f be such that $f \circ h$ is in the domain of the generator of X. Then:

$$f(Y_t) - f(Y_0) = (f \circ h)(X_{a_t}) - (f \circ h)(X_0) = M_{a_t} + \int_0^{a_t} L^X(f \circ h)(X_u)du,$$

where M is a local martingale.

Implementing the time change $u = a_s$, we obtain:

$$f(Y_t) - f(Y_0) = M_{a_t} + \int_0^t L^X(f \circ h)(X_{a_s})da_s$$

$$= M_{a_t} + \int_0^t L^X(f \circ h)(X_{a_s})\frac{ds}{k(X_{a_s})}$$

$$= M_{a_t} + \int_0^t L^X(f \circ h)(h^{-1}(Y_s))\frac{ds}{k(h^{-1}(Y_s))}.$$

The process $f(Y_t) - f(Y_0) - \int_0^t L^X(f \circ h)(h^{-1}(Y_s))ds/k(h^{-1}(Y_s))$ is thus a local martingale, which enables to identify $L^Y(f)$.

Consequence. Denoting $f_m(x) \equiv x^m$, we note that the Lévy exponent ψ associated with a semi-stable Markov process can be calculated via the formula

$$L f_m(x) = x^{m-1}\psi(m).$$

Proof of Proposition 4.1. We prove the formula which gives the expression for the semigroup; the formula giving the generator follows immediately from that.

Let us first show that:

$$T_t^\alpha f(x) = f(x)e^{-t/2x} + \int_0^{t/2x} e^{-z} E_0[f(x + \mu\ell(\tau_{t-2xz}^{\mu,x}))]dz$$

where $\tau_t^{\mu,a} \equiv \inf\{u; \ell_u^a(X^\mu) > t\}$ denotes the right-continuous reciprocal of the local time at the level a of the μ-process.

We recall Remark 6 following Proposition 4.2, which states that

$$T_t^\alpha f(x) = E_x[f(|B(\tau_t^{\mu,x})|)]$$

where $\tau_t^{\mu,x}$ is the reciprocal of the local time at the level x of the μ-process constructed from B (starting at x).

For the calculation of the right-hand side of the above equality, two cases arise.

• Either $\tau_t^{\mu,x} < T_0(B)$, that is $t < \ell_{T_0}^x(B)$, where $T_a(B)$ denotes the first passage time to a by B. Then, using the Ray–Knight theorems on local times of real-valued Brownian motion:

$$E_x[f(|B(\tau_t^{\mu,x})|)]1_{\{\tau_t^{\mu,x}<T_0(B)\}}] = f(x)P[\ell_{T_0}^x(B) > t] = f(x)\mathbb{P}[Z_1 > t/2x]$$
$$= f(x)e^{-t/2x}.$$

• Or $\tau_t^{\mu,x} \geq T_0(B)$. Then on this set:

$$\tau_t^{\mu,x} = T_0 + \inf\{s; \ell_{s+T_0}^x(X^\mu) > t\} = T_0 + \inf\{s; \ell_{T_0}^x(X^\mu) + \ell_s^x(\hat{X}) > t\}$$

where $\hat{X}$ is the μ-process associated with the Brownian motion independent of $\mathcal{F}_{T_0}$ and starting at zero, $\hat{B}_t = B(t+T_0)$. Since $\ell_{T_0}^x(X^\mu) = \ell_{T_0}^x(B)$, we have, with obvious notation: $\tau_t^{\mu,x} = T_0 + \hat{\tau}^{\mu,x}(t - \ell_{T_0}^x(B))$. It follows that:

$$E_x[f(|B(\tau_t^{\mu,x})|)]1_{\{\tau_t^{\mu,x}\geq T_0(B)\}}] = E_x[f(|\hat{B}(\hat{\tau}^{\mu,x}(t - \ell_{T_0}^x(B)))|)]$$
$$= \int_0^t \mathbb{P}_x[\ell_{T_0}^x(B) \in dz]E_0[f(|B(\tau_{t-z}^{\mu,x})|)]$$
$$= \int_0^t \mathbb{P}[2xZ_1 \in dz]E_0[f(x + \mu\ell(\tau_{t-z}^{\mu,x}))]$$

because $X^\mu + \mu\ell = |B|$.

Grouping together the two expressions, we find after a change of variable:

$$T_t^\alpha f(x) = f(x)e^{-t/2x} + \int_0^{t/2x} e^{-z}E_0[f(x + \mu\ell(\tau_{t-2xz}^{\mu,x}))]dz.$$

It then remains to identify the distribution of $\ell_{\tau_t^{\mu,a}}$ for $a > 0$, which is hard to access. For this, we adapt the proof of a Ray–Knight theorem for the μ-process given in [20] (Chapter 9, Theorem 9.1, pages 118ff), taking into account what happens at the point a and no longer at zero.

$$\int_0^{+\infty} e^{-t\theta^2/2} E[q(\ell_{\tau_t^{\mu,a}})]dt$$

$$= 2 \int_0^{+\infty} q(s) E_0 \left[\exp\left(-\frac{\theta^2}{2}\ell_{\tau_s^a}(X^\mu)\right)\right] E_{a+\mu s}\left[\exp\left(-\frac{\theta^2}{2}\ell_{T_0}^{a+\mu s}(B)\right)\right] ds$$

$$= \int_0^{+\infty} q(s)(1+a\theta^2)^{1/\mu}(1+\theta^2(a+\mu s))^{-1/\mu}ds$$

using the Ray–Knight theorems which describe, in terms of Bessel squares, the processes of the local times of the μ-process at time τ_t, on the one hand ([20], Chapter 3, Theorem 3.4, page 33) and of the Brownian motion starting at a and at its first passage time to zero, on the other hand ([20], Chapter 3, RK2 page 28).

The proof is then completed by inverting the Laplace transform in $\theta^2/2$.

Acknowledgments

We are grateful to the referee for suggestions which have enabled us to write a better presentation of our results.

References

1. Biane, P. and Yor, M. (1989). Sur la loi des temps locaux browniens pris en un temps exponentiel indépendant. Séminaire de Probabilités XXIII. *Lecture Notes in Maths,* **1321**, 454–466. Springer
2. Bingham, N.H., Goldie, C.M. and Teugels, J.L. (1987). Regular Variation. Encyclopedia of Mathematics and its Applications, vol. **27**. Cambridge University Press
3. Blumenthal, R.M. and Getoor, R.K. (1968). *Markov Processes and Potential Theory.* Academic Press
4. Carmona, P., Petit, F. and Yor, M. (1994). Some extensions of the arc sine law as partial consequences of the scaling property of Brownian motion. *Prob. Th. Rel. Fields,* **100**, 1–29
5. Dufresne, D. (1989). Weak convergence of random growth process with application to insurance. *Insur. Math. Econ.,* **8**, 187–201
6. Dufresne, D. (1990). The distribution of perpetuity with applications to risk theory and pension funding. *Scand. Actuarial J.,* 39–79
7. Dynkin, E.B. (1965). *Markov Processes,* vol. **I**, Springer
8. Feller, W. (1968). *An Introduction to Probability Theory and its Applications,* vol. **I**, 3rd edn, Wiley
9. Jeulin, T. and Yor, M. (1981). Sur les distributions de certaines fonctionnelles du mouvement brownien. Séminaire de Probabilités XV. *Lecture Notes in Maths.,* **850**, 210–226. Springer

10. Kiu, S.W. (1980). Semi-stable Markov processes in $\mathbb{R}^n$. *Stochastic Processes Appl.*, **10**, 183–191
11. Lamperti, J.W. (1967). Continuous state branching processes. *Bull. Am. Math. Soc.*, **73**, 382–386
12. Lamperti, J.W. (1972). Semi-stable Markov processes. *Z. Wahrscheinlichkeit*, **22**, 205–225
13. Le Gall, J.F. and Yor, M. (1986). Excursions Browniennes et carrés de Processus de Bessel. *C. R. Acad. Sci., Paris, Sér. I*, **303**, 73–76
14. Pazy, A. (1983). *Semigroups of Linear Operators and Applications to Partial Differential Equations*, Applied Mathematical Sciences, vol. **44**, Springer
15. Pitman, J. and Rogers, L. (1981). Markov functions. *Ann. Probab.*, **9** (4), 573–582
16. Revuz, D. and Yor, M. (1991). *Continuous Martingales and Brownian Motion.* Springer
17. Vuolle-Apiala, J. (1989). Times changes of a self-similar Markov process. *Ann. Inst. Henri Poincaré, Probab. Stat.*, **25** (4), 581–587
18. Watanabe, S. (1980). A limit theorem for sums of i.i.d. random variables with slowly varying tail probability. In 'Multivariate Analysis', vol. **5**, Krishnaia, P.R. (ed), pp. 249–261. (Amsterdam: North-Holland)
19. Yor, M. (1989). Une extension markovienne de l'algèbre des lois béta-gamma. *C. R. Acad. Sci., Paris, Sér. I*, **308**, 257–260
20. Yor, M. (1992). Some Aspects of Brownian Motion. Part I: Some Special Functionals. *Lectures in Maths.*, ETH Zurich. (Basel: Birkhäuser)
21. Yor, M. (1992). Sur les lois des fonctionnelles exponentielles du mouvement brownien, considérées en certains instants aléatoires. *C. R. Acad. Sci., Paris, Sér. I*, **314**, 951–956. **Paper [4] in this volume**
22. Yor, M. (1992). Intertwinings of Bessel Processes. Tech. Report, University of California at Berkeley [*See the Postscript below*]

Postscript #8

The results obtained in this paper hinge mainly on Lamperti's representation of the exponential of a Lévy process. They are further developed in:

P. Carmona, F. Petit, M. Yor (1997): "*On the distribution and asymptotic results for exponential functionals of Lévy processes*". In: "Exponential Functionals and Principal Values related to Brownian motion", Bibl. Rev. Math. Ib. Amer., M. Yor (ed), p. 73–121

which became the final version of the article [22] mentioned in the above references. A further article (∗) by Carmona-Petit-Yor, mentioned in Postscript #1, d), aims in particular at covering the literature on this subject, from 1993 till 2000. The works of Gjessing, Paulsen, Urbanik... are mentioned in (∗), as well as in the paper by Bertoin-Yor quoted in Postscript #1, b).

On Some Exponential-integral Functionals
of Bessel Processes

Mathematical Finance, Vol. **3**, No. 2 (April 1993), 231–240

Abstract. This paper studies the moments of some exponential-integral functionals of Bessel processes, which are of interest in some questions of mathematical finance, including the valuation of perpetuities and Asian options.

1.   Introduction

The aim of this paper is to exhibit some properties of the laws of the following random variables:

$$Z_\xi = \int_0^\infty dt \, \exp\left(-\xi \int_0^t ds \, R_s^\gamma\right),\tag{1.1}$$

where $(R_s; s \geq 0)^{(1)}$ denotes a Bessel process with dimension $\delta = 2(1 + \nu)$, starting from r. To simplify the discussion, we shall assume that $\nu \geq 0$, i.e., $\delta \geq 2$; moreover, both parameters ξ and γ are assumed to be strictly positive.

In Section 2, we give another representation of Z_ξ, using the property that a power of a Bessel process is another Bessel process time-changed. This second representation allows one to obtain, in Section 3, an integral expression for the first moment of Z_ξ.

Then we prove in Section 4 that Z_ξ has moments of all orders; moreover, it will be shown that the law of Z_ξ is determined by its moments, since Carleman's criterion applies, and that this law has moments of exponential type (see Theorem 4.1 for a precise statement).

In Section 5, a number of variants are presented, in particular with the Bessel process $(R_s, s \geq 0)^{(1)}$ being replaced by the norm of a multidimensional Ornstein-Uhlenbeck process.

2.   A Second Representation of Z_ξ

In the sequel we shall use, in an essential manner, the fact that, if $1/p + 1/q = 1$, there exists a Bessel process $(R_{\nu q}(t); t \geq 0)$ with index $\mu \equiv \nu q$, starting from $\rho = qr^{1/q}$, such that

$$qR_\nu^{1/q}(t) = R_{\nu q}\left(\int_0^t ds \, R_\nu^{-2/p}(s)\right), \qquad t \geq 0\tag{2.1}$$

$^{(1)}$ In the rest of the text, we shall write $(R_\nu(s), s \geq 0)$ to indicate the dependency with respect to ν.

(see Biane and Yor 1987, Lemme 3.1, p. 40, or Revuz and Yor 1991, p. 416). We may now represent Z_ξ as follows.

Proposition 2.1. *Using the above notation, we have*

$$Z_\xi = q^{\gamma q} \int_0^\infty \frac{dt \exp(-\xi t)}{(R_{\nu q}(t))^{\gamma q}}, \qquad where \quad \frac{1}{q} = 1 + \frac{\gamma}{2}. \tag{2.2}$$

Proof. We shall apply (2.1) with $p = -2/\gamma$. We then obtain $1/q = 1 + \gamma/2$, and from (2.1) we deduce, if we let

$$A_t = \inf \left\{ u : \int_0^u ds\ R_\nu^\gamma(s) > t \right\},$$

that

$$R_\nu(A_t) = \left(\frac{1}{q} R_{\nu q}(t) \right)^q, \qquad t \geq 0.$$

Now, we may write Z_ξ as

$$Z_\xi = \int_0^\infty dA_t \exp(-\xi t) = \int_0^\infty \frac{dt}{R_\nu^\gamma(A_t)} \exp(-\xi t) = \int_0^\infty \frac{dt \exp(-\xi t)}{(R_{\nu q}(t)/q)^{\gamma q}},$$

which gives (2.2). $\square$

In the sequel, we shall also use the following remark: the square, starting at x, of a Bessel process with dimension δ, which we denote for now by $(X_x^{(\delta)}(t); t \geq 0)$ may be obtained by adding two independent squares of Bessel processes, namely, $(X_x^{(0)}(t); t \geq 0)$ and $(X_0^{(\delta)}(t); t \geq 0)$; that is,

$$(X_x^{(\delta)}(t); t \geq 0) \overset{(\text{law})}{=} (X_x^{(0)}(t) + X_0^{(\delta)}(t); t \geq 0). \tag{2.3}$$

We shall use this property in the following way: if $F : \mathbb{R}_+ \to \mathbb{R}_+$ is a decreasing function, then

$$E\left[F\left(\int_0^t ds\ (X_x^{(\delta)}(s))^m \right) \right] \leq E\left[F\left(\int_0^t ds\ (X_0^{(\delta)}(s))^m \right) \right]. \tag{2.4}$$

3. An Integral Formula for the First Moment of Z_ξ

We need the following

Lemma 3.1. *(see Yor 1991, Lemme 2).* Let $\alpha > 0$ and $\mu \geq 0$. Then, denoting by $Q_z^{(\mu)}$ the law of the square, starting from z, of a Bessel process with index μ, we have

$$Q_z^{(\mu)} \left(\frac{1}{(X_t)^\alpha} \right) = \frac{1}{\Gamma(\alpha)} \int_0^1 dh\ h^{\alpha-1}(1-h)^{\mu-\alpha} \frac{1}{(2t)^\alpha} \exp\left(-\frac{zh}{2t} \right). \tag{3.1}$$

In order to apply Lemma 3.1 to the computation of the first moment of Z_ξ, we take

$$\mu = \nu q; \quad \alpha = \frac{\gamma q}{2} \equiv \frac{\gamma}{2+\gamma}; \quad z = \rho^2 = (qr^{1/q})^2.$$

We then deduce from (2.2) and (3.1) that

$$E[Z_\xi] = \frac{q^{\gamma q}}{\Gamma(\alpha)} \int_0^1 dh \, h^{\alpha-1}(1-h)^{\mu-\alpha} \int_0^\infty \frac{dt}{(2t)^\alpha} \exp\left(-\frac{\rho^2 h}{2t} - \xi t\right). \quad (3.2)$$

Now, the integral in dt may be represented in terms of McDonald's functions K_θ; indeed, we recall (see, e.g., Lebedev 1974, (5.10.25), p. 119) the classical integral representation formula

$$K_\theta(z) = \frac{1}{2} \int_0^\infty \frac{du}{u^{\theta+1}} \exp -\frac{z}{2}\left(u + \frac{1}{u}\right),$$

from which it follows

$$\int_0^\infty \frac{d\nu}{\nu^\alpha} \exp -\frac{1}{2}\left(b^2\nu + \frac{c^2}{\nu}\right) = \left(\frac{b}{c}\right)^{\alpha-1} 2K_{\alpha-1}(bc). \quad (3.3)$$

We may now state and prove

Proposition 3.2. *Take* $\xi = b^2/2$. *Then* $E[Z_{b^2/2}]$ *admits the integral representation*

$$E[Z_{b^2/2}] = \frac{q^{\gamma q}}{\Gamma(\alpha)} \left(\frac{2\rho}{b}\right)^{1-\alpha} \int_0^1 dh \, h^{(\alpha-1)/2}(1-h)^{\mu-\alpha} K_{1-\alpha}\left(\rho b\sqrt{h}\right). \quad (3.4)$$

Proof.   Formula (3.4) follows from (3.2) and (3.3), with $c = \rho\sqrt{h}$.   □

Letting r, hence ρ, go to 0 in (3.4), we obtain the following result.

Corollary 3.1. *We let* $\xi = b^2/2$ *and* $\alpha = \gamma/(2+\gamma)$ (*so that, with our previous notation, we have* $p = -2/\gamma$ *and* $q = 1 - \alpha = 2/(2+\gamma)$). *Then*

(a) $$E_0(Z_\xi) = (1-\alpha)^{\gamma(1-\alpha)} \left(\frac{2}{b}\right)^{2(1-\alpha)} \frac{\Gamma(1-\alpha)\Gamma(\mu+1-\alpha)}{\Gamma(\mu+1)}$$

$$= \frac{C}{\xi^{1-\alpha}},$$

where

$$C = (2(1-\alpha)^\gamma)^{1-\alpha} \frac{\Gamma(1-\alpha)\Gamma(\mu+1-\alpha)}{\Gamma(\mu+1)}.$$

(b) *If we let* $X = \int_0^1 ds\, (R_\nu(s))^\gamma$, *we have*

$$E_0(Z_\xi) = E_0\left(\frac{1}{(\xi X)^{1-\alpha}}\right)\Gamma(2-\alpha).$$

(c) *We have the identity in law*

$$X^{1-\alpha} \overset{(\text{law})}{=} (1-\alpha)^{-\gamma(1-\alpha)}\left\{\int_0^1 du\, (R_\mu(u))^{-\gamma(1-\alpha)}\right\}^{-1}, \tag{3.5}$$

from which it follows that

$$E_0(X^{-(1-\alpha)}) = (1-\alpha)^{\gamma(1-\alpha)}\frac{1}{(1-\gamma(1-\alpha))}E_0((R_\mu(1))^{-\gamma(1-\alpha)}).$$

Proof. The equality in (a) follows from (3.4), and the asymptotic result

$$K_\theta(z) \sim \frac{\Gamma(\theta)}{2}\left(\frac{z}{2}\right)^{-\theta}, \qquad \text{as } z \to 0.$$

The equality in (b) follows from the definition of Z_ξ given in (1.1) and the scaling property for $(R_\nu(s), s \geq 0)$ when $R_\nu(0) = 0$.

The identity in law (3.5) is a consequence of the identity in law (2.1) and of the scaling property of both R_ν and R_μ, starting at 0. This identity in law was already presented in Biane and Yor (1987, Corollaire (3.2)). The final equality is a consequence of (3.5) and, again, of the scaling property of R_μ. □

For $\gamma = 2$, it is possible to obtain a formula equivalent to (3.4) by taking a route different from that above, which relied essentially on (2.4). Indeed, in this case, we shall use the following formula, which is close to Lévy's formula for the stochastic area of Brownian motion:[1]

$$E_r\left[\exp\left(-\frac{b^2}{2}\int_0^t ds\, R_s^2\right)\right] = (\text{ch}(bt))^{-\delta/2}\exp\left(-\frac{r^2b}{2}\text{th}(bt)\right) \tag{3.6}$$

(see Pitman and Yor 1982, (2.k), or Revuz and Yor 1991, Corollary (1.8), p. 414). Consequently, we obtain

$$E[Z_{b^2/2}] = \int_0^\infty \frac{dt}{(\text{ch}(bt))^{\delta/2}}\exp\left(-\frac{r^2b}{2}\text{th}(bt)\right)$$

$$= \frac{1}{b}\int_0^\infty \frac{ds}{(\text{ch}(s))^{\delta/2}}\exp\left(-\frac{r^2b}{2}\text{th}(s)\right).$$

By making the change of variables $u = \text{th}(s)$, we have

$$E[Z_{b^2/2}] = \frac{1}{b}\int_0^1 du(1-u^2)^{(\nu-1)/2}\exp\left(-\frac{r^2bu}{2}\right). \tag{3.7}$$

---

[1] The French abbreviations ch and th for cosh and tanh are used throughout.

Now, we remark that (3.7) coincides with (3.4), since for $\gamma = 2$ we have $\alpha = q = 1/2$ and $K_{1/2}(x) = (\pi/2x)^{1/2} \exp(-x)$ (see, e.g., Lebedev 1972, (5.8.5), p. 112). Hence, (3.4) reads

$$E[Z_{b^2/2}] = \frac{1}{2b} \int_0^1 \frac{dh}{\sqrt{h}} (1-h)^{(\nu-1)/2} \exp\left(-\frac{r^2 b}{2}\sqrt{h}\right),$$

which finally gives (3.7) if we let $h = u^2$.

4. Some Estimates of the Moments of Z_ξ

We use Z_ξ instead of $Z_{b^2/2}$. The following statements are the main results of this paper.

Theorem 4.1. *Define $Z_\xi = \int_0^\infty dt \exp(-\xi \int_0^t ds R_s^\gamma)$, where $(R_t, t \geq 0)$ denotes a Bessel process with index $\nu \geq 0$, starting from $r \geq 0$. Define $\alpha = \gamma/(2+\gamma)$. Then*

  (i) $E_0(Z_\xi) = C/\xi^{1-\alpha}$, where C is given in statement (a) of Corollary 3.1.
  (ii) $E_r((Z_\xi)^n) \leq (n!)^\alpha C^n/\xi^{n(1-\alpha)}$.
  (iii) *If we let $m_n = E_r[(Z_\xi)^n]$, then Carleman's criterion*

$$\sum_{n=1}^\infty \frac{1}{(m_n)^{1/2n}} = \infty$$

is satisfied. Hence, the law of Z_ξ is determined by its moments $(m_n; n \in \mathbb{N})$.
  (iv) *Let p be an integer such that $p \leq 1/\alpha \equiv (2+\gamma)/\gamma$.*
Then there exists $a > 0$ such that $E[\exp(aZ^p)] < \infty$. Hence, for $\gamma = 2$ there exists $a > 0$ such that $E[\exp(aZ^2)] < \infty$.

Proof. (1) Equality (i) has already been stated in Corollary 3.1.

(2) To obtain inequality (ii), we develop $E_r((Z_\xi)^n)$ as follows:

$$E_r((Z_\xi)^n) = n! E\left[\int_0^\infty dt_1 \exp(-\xi I_{(0,t_1)}) \int_{t_1}^\infty dt_2 \right.$$
$$\left. \cdots \exp(-\xi I_{(0,t_2)}) \cdots \int_{t_{n-1}}^\infty dt_n \exp(-\xi I_{(0,t_n)})\right]$$

where

$$I_{(s,t)} = \int_s^t du \exp(-\xi R_u^\gamma).$$

Decomposing, for every $k > 1, I_{(0,t_k)}$ into $I_{(0,t_k)} = I_{(0,t_{k-1})} + I_{(t_{k-1},t_k)}$, we obtain

$$E_r[(Z_\xi)^n] = n! E \left[\int_0^\infty dt_1 \exp(-n\xi I_{(0,t_1)}) \int_{t_1}^\infty dt_2 \exp(-(n-1)\xi I_{(t_1,t_2)}) \right.$$

$$\left. \cdots \int_{t_{n-2}}^\infty dt_{n-1} \exp(-2\xi I_{(t_{n-2},t_{n-1})}) \int_{t_{n-1}}^\infty dt_n \exp(-\xi I_{(t_{n-1},t_n)}) \right].$$

Going from right to left, we apply the Markov property n times and we replace successively $R_{t_{n-1}}$ by 0, then $R_{t_{n-2}}$ by 0, and so on, thanks to (2.4). With this procedure, we obtain

$$E_r((Z_\xi)^n) \leq n! E_0 \left[\int_0^\infty dt\, e^{-n\xi I_{(0,t)}} \right] E_0 \left[\int_0^\infty dt\, e^{-(n-1)\xi I_{(0,t)}} \right]$$

$$E_0 \left[\int_0^\infty dt\, e^{-\xi I_{(0,t)}} \right] \leq n! \prod_{k=1}^n (E_0(Z_{k\xi})).$$

Inequality (ii) now follows from (i).

(3) We now remark that, as a consequence of (ii), we have

$$(m_n)^{1/2n} \leq D_\xi (n!)^{\alpha/2n} \leq D_\xi\, n^{\alpha/2},$$

where we may take $D_\xi = (C/\xi^{1-\alpha})^{1/2}$. Consequently, we have

$$\sum_n \frac{1}{(m_n)^{1/2n}} \geq \sum_n \frac{1}{D_\xi} \frac{1}{n^{\alpha/2}} = \infty, \qquad \text{since } \alpha < 1.$$

(4) Using inequality (ii) again, we obtain

$$E[\exp(aZ^p)] = \sum_{n=0}^\infty \frac{1}{n!} E_r[(aZ^p)^n] \leq \sum_{n=0}^\infty \frac{a^n}{n!} \frac{((pn)!)^\alpha C^{pn}}{\xi^{pn(1-\alpha)}}$$

$$\leq \sum_n (F_{a,\xi})^n \frac{((pn)!)^\alpha}{n!},$$

where $F_{a,\xi} = C^p a/\xi^{p(1-\alpha)}$. Applying Stirling's formula, we obtain

$$\frac{((pn)!)^\alpha}{n!} \simeq \left(\frac{p^{p\alpha}}{e^{-(1-p\alpha)}} \right)^n n^{(p\alpha-1)n} p^{\alpha/2}\, n^{(\alpha-1)/2} (2\pi)^{\frac{\alpha-1}{2}}$$

Hence, if $p\alpha \leq 1$, the series is convergent when a is sufficiently small in the case $p\alpha = 1$ and for every a if $p\alpha < 1$. □

As we now know that Z_ξ has moments of all orders and that these moments determine the law of Z_ξ, the question of the actual computation of these moments arises naturally. The second representation of Z_ξ, obtained in Proposition 2.1, makes it possible to express those moments in terms of iterates of the resolvent of the Bessel process R_μ, with index $\mu = \nu q$.

Indeed, (2.2) may be written as

$$Z_\xi = \int_0^\infty dt\, f(X_t) \exp(-\xi t), \qquad (4.1)$$

where $(X_t = R_\mu(t), t \geq 0)$ is the Bessel process with index μ, starting from $\rho \equiv q r^{1/q}$, and $f(x) = (x/q)^{\gamma q}$.

Now, a variant of the computations made in (2) of the proof of Theorem 4.1 shows that

$$E_r[(Z_\xi)^n] = n!\, U_{n\xi}(f(U_{(n-1)\xi}\,(f(U_{(n-2)\xi}\,(f(\ldots U_\xi(f)\ldots)\ldots))))\,(\rho). \quad (4.2)$$

Moreover, $U_\xi(f)(\rho)$ is given by (3.4), in which we replace b by $\sqrt{2\xi}$.

5.   Some Complements

It seems that either from the point of view of potential applications (in mathematical finance; perpetuities, for example) or in order to get explicit, simple enough, formulae, the case $\gamma = 2$ is the most important to consider. In this case, if we use a more complete formula than (3.7), namely $(2.k)$ in Pitman and Yor (1982), which is

$$E_r^{(\delta)}\left[\exp\left(-\frac{\alpha}{2}R_t^2 - \frac{\beta^2}{2}\int_0^t ds\, R_s^2\right)\right]$$
$$= (\mathrm{ch}(\beta t) + \frac{\alpha}{\beta}\mathrm{sh}(\beta t))^{-\delta/2} \exp -\frac{r^2\beta}{2}\left\{\frac{(\alpha/\beta) + \mathrm{th}(\beta t)}{1 + (\alpha/\beta)\,\mathrm{th}(\beta t)}\right\}, \qquad (5.1)$$

it is easy enough to obtain a closed (integral) formula for the second moment of Z_ξ.

Indeed, using the notation $I_{(0,t)}$, as in the proof of Theorem 4.1, we have

$$E_r[(Z_\xi)^2] = 2E_r\left[\int_0^\infty dt\, \exp(-\xi I_{(0,t)})\int_t^\infty ds\, \exp(-\xi I_{(0,s)})\right]$$
$$= 2E_r\left[\int_0^\infty dt\, \exp(-2\xi I_{(0,t)})\, E_{R_t}\left(\int_0^\infty ds\, \exp(-\xi I_{(0,s)})\right)\right]$$
$$= 2E_r\left[\int_0^\infty dt\, \exp(-2\xi I_{(0,t)})\frac{1}{b}\int_0^1 dx\,(1-x^2)^{(\nu-1)/2}\exp\left(-\frac{R_t^2 bx}{2}\right)\right] \quad (5.2)$$
$$= \frac{2}{b}\int_0^1 dx\,(1-x^2)^{(\nu-1)/2}\int_0^\infty dt\, E_r\left[\exp\left(-\frac{R_t^2(bx)}{2} - 2\xi I_{(0,t)}\right)\right],$$

where we used (3.7) in (5.2).

Using now (5.1) with $\alpha = bx$ and $\beta = 2b$, we obtain

$$E_r[(Z_\xi)^2] = \frac{2}{b} \int_0^1 dx\, (1-x^2)^{(\nu-1)/2} \int_0^\infty dt\, (\mathrm{ch}(2bt) + \frac{x}{2}\,\mathrm{sh}(2bt))^{-\delta/2}$$
$$\cdots \exp{-r^2 b \left\{ \frac{x/2 + \mathrm{th}(2\beta t)}{1 + (x/2)\,\mathrm{th}(2\beta t)} \right\}}.$$

Making the change of variables $y = \mathrm{th}(2bt)$, we finally obtain

$$E_r[(Z_\xi)^2]$$
$$= \frac{1}{b^2} \int_0^1 dx\, (1-x^2)^{(\nu-1)/2} \int_0^1 dy\, \frac{(1-y^2)^{(\nu-1)/2}}{(1+xy/2)^{\delta/2}} \exp{-r^2 b \left\{ \frac{y + x/2}{1 + xy/2} \right\}}$$
$$\tag{5.3}$$

(we recall the relation $\delta = 2(\nu + 1)$ between the index ν and the dimension δ).

Using (5.1) jointly with Girsanov's theorem allows to extend some of the above computations to the norm of a δ-dimensional Ornstein-Uhlenbeck process, with parameter θ, that is, the norm of a process $(U_t, t \geq 0)$ which is the solution of

$$U_t = u + B_t - \theta \int_0^t U_s\, ds, \quad t \geq 0,$$

where $(B_t, t \geq 0)$ is a δ-dimensional Brownian motion starting from 0.

Girsanov's theorem gives

$$P_r^{(\theta)}|_{\mathscr{R}_t} = \exp\left\{ -\frac{\theta}{2}(R_t^2 - r^2 - \delta t) - \frac{\theta^2}{2} \int_0^t ds\, R_s^2 \right\} \cdot P_r|_{\mathscr{R}_t}, \tag{5.4}$$

where $\mathscr{R}_t = \sigma(R_s, s \leq t)$ and $P_r^{(\theta)}$ denotes the law of $(R_t^{(\theta)} = |U_t|, t \geq 0)$ (in fact, (5.4) is also valid for any $\delta > 0$; see Pitman and Yor (1982) for more details).

If we now define

$$Z_\xi^{(\theta)} = \int_0^\infty dt \exp\left(-\xi \int_0^t ds\, (R_s^{(\theta)})^2 \right),$$

we may compute

$$E_r[Z_\xi^{(\theta)}] = \int_0^\infty dt\, E_r^{(\theta)}\left[\exp{-\xi \int_0^t ds\, R_s^2} \right]$$
$$= \exp\left(\frac{\theta r^2}{2} \right) \int_0^\infty dt \exp\left(\frac{\theta \delta t}{2} \right) E_r\left[\exp\left(-\frac{\theta R_t^2}{2} - \frac{b^2 + \theta^2}{2} \int_0^t ds\, R_s^2 \right) \right]$$

thanks to (5.4).

Using now (5.1) with $\alpha = \theta$ and $\beta \overset{\text{def}}{=} (b^2 + \theta^2)^{1/2}$, we obtain

$$E_r(Z_\xi^{(\theta)}) = \exp\left(\frac{\theta r^2}{2}\right) \int_0^\infty dt \exp\left(\frac{\theta \delta t}{2}\right) \left(\text{ch}(\beta t) + \frac{\theta}{\beta}\text{sh}(\beta t)\right)^{-\delta/2}$$
$$\cdots \exp -\frac{r^2 \beta}{2} \left\{\frac{(\theta/\beta) + \text{th}(\beta t)}{1 + (\theta/\beta)\text{th}(\beta t)}\right\}.$$

It now remains to make the change of variables $u = \text{th}(\beta t)$ to obtain a somewhat simpler integral expression.

Another potentially important complement is that a large part of the above results extends to the more general class of random variables

$$Z'_\xi = \int_0^\infty \frac{dt}{R_t^\beta} \exp\left(-\xi \int_0^t ds\ R_s^\gamma\right),$$

where $(R_s, s \geq 0)$ again denotes a Bessel process with dimension $\delta = 2(1+\nu)$, starting from r.

The analog of (2.2) is

$$Z'_\xi = \int_0^\infty \frac{dt \exp(-\xi t)}{(R_{\nu q}(t)/q)^{(\beta+\gamma)q}}, \qquad \text{where } \frac{1}{q} = 1 + \frac{\gamma}{2}. \tag{5.5}$$

The above quantities may also be expressed in terms of exponentials of Brownian motion with drift. Indeed, we recall that[1]

$$\exp(B_t + \nu t) = R_\nu(A_t^{(\nu)}), \qquad \text{where } A_t^{(\nu)} = \int_0^t ds\ \exp 2(B_s + \nu s)$$

(this relation serves as a key argument in Yor (1991) and related papers). Precisely, if $R_\nu(0) = 1$, we have, in the previous notation,

$$Z'_\xi \equiv \int_0^\infty \frac{dt}{(R_\nu(t))^\beta} \exp\left(-\xi \int_0^t ds\ R_\nu^\gamma(s)\right)$$
$$= \int_0^\infty \frac{dA_u^{(\nu)}}{(R_\nu(A_u^{(\nu)}))^\beta} \exp\left(-\xi \int_0^u dA_h^{(\nu)}\ R_\nu^\gamma(A_h^{(\nu)})\right)$$
$$= \int_0^\infty du\ \exp[(2-\beta)(B_u + \nu u)] \exp -\xi \int_0^u dh\ \exp[(2+\gamma)(B_h + \nu h)].$$

Thus, the variables Z'_ξ appear as particular cases of the five-parameter family of random variables

$$\int_0^\infty du\ \exp(-au + bB_u) \exp\left(-c \int_0^u ds\ \exp(pB_s + qs)\right).$$

---

[1] This is Lamperti's relation; see Postscript #1.

References

Biane, Ph. and Yor, M. (1987). Valeurs Principales Associées aux Temps Locaux Browniens. *Bull. Sci. Math.*, **111**, 23–101

Delbaen, F. (1993). Consols in the CIR Model. *Math. Finance*, **3**, 125–134

Geman, H. and Yor, M. (1993). Bessel Processes, Asian Options, and Perpetuities. *Math. Finance*, **3**, 349–375. **Paper [5] in this book**

Lebedev, N. (1972). *Special Functions and Their Applications.* New York: Dover

Pitman, J. and Yor, M. (1982). A Decomposition of Bessel Bridges, *Zeit. Wahrsch. Geb.*, **59**, 425–457

Revuz, D. and Yor, M. (1991). *Continuous Martingales and Brownian Motion.* New York: Springer

Yor, M. (1991). Factorisation de Fonctionnelles Exponentielles du Mouvement Brownien et de Certains Processus à Accroissements Indépendants, preprint, ETH, Zürich *(See the Postscript below)*

Acknowledgment

I am grateful to F. Delbaen and H. Geman for providing the original motivation for this study. They were interested in the variables Z_ξ and some variants of them in connection with questions of mathematical finance, especially the valuation of perpetuities (see Delbaen 1993 and Geman and Yor 1993).

Postscript #9

The contents of the preprint in the last reference above are found in paper [6] in this book.

Exponential Functionals of Brownian Motion and Disordered Systems

J. Appl. Prob. **35** (1998), 255–271
(with Alain Comtet and Cécile Monthus)

Abstract. The paper deals with exponential functionals of the linear Brownian motion which arise in different contexts, such as continuous time finance models and one-dimensional disordered models. We study some properties of these exponential functionals in relation with the problem of a particle coupled to a heat bath in a Wiener potential. Explicit expressions for the distribution of the free energy are presented.

1. Introduction

Consider a linear Brownian motion $(B_s, s \geq 0, B_0 = 0)$, and a given drift μ. Exponential functionals of the following form,

$$A_t^{(\mu)} = \int_0^t ds \exp[-2(B_s + \mu s)], \tag{1}$$

have recently been a subject of common interest for mathematicians and for physicists.

Some recent mathematical studies have been partly motivated by continuous time finance models, in which most stock price dynamics are assumed to be driven by the exponential of a Brownian motion with drift [12, 14]. In such studies, the following representation

$$\exp[-(B_t + \mu t)] = R\left(\int_0^t ds \exp[-2(B_s + \mu s)]\right), \tag{2}$$

where $(R(u), u \geq 0)$ is a Bessel process, i.e. an element of an important class of diffusions, exhibits the importance of the functional $A_t^{(\mu)}$. Formula (2) is a particular instance of the Lamperti relation which expresses $(\exp(\xi_t), t \geq 0)$ as

$$e^{\xi_t} = X\left(\int_0^t ds \, e^{2\xi_s}\right), \tag{3}$$

where $(\xi_t, t \geq 0)$ is a Lévy process and $(X(u), u \geq 0)$ is a semi-stable Markov process (see [9, 10] for some applications, partly in mathematical finance).

In physics, these exponential functionals play a central role in the context of one-dimensional classical diffusion in a random environment. The random variable $A_\infty^{(\mu)}$ can indeed be interpreted as a trapping time. Its probability distribution controls the anomalous diffusive behaviors of the particle at large time in a infinite sample [5, 4, 15]. The distribution of $A_L^{(\mu)}$ occurs when studying the maximum reached by the Brownian particle in a drifted Brownian potential [17]. The functional $A_L^{(\mu)}$ also arises in the study of the transport properties of disordered samples of finite length L [21, 24, 25]. In fact, $A_L^{(\mu)}$ represents the continuous space analogue of the random series introduced by Kesten et $al.$ [19] for the so called 'random random walk'. This random series is generated by a linear recurrence relation with random coefficients and therefore constitutes a discrete random multiplicative stochastic process. $A_t^{(\mu)}$ also represents the very simple case of a continuous random multiplicative stochastic process [11] which may be related to hyperbolic Brownian motion [11, 31].

In this article, we discuss some properties of these functionals, concentrating mostly on the mean-value $E\left(\ln A_t^{(\mu)}\right)$, in relation with another physical interpretation inspired by the statistical mechanics of disordered systems. In these systems, the partition function Z is a functional which depends on a set of 'quenched' random couplings. In order to obtain the thermodynamic properties of the system, one has to compute the average over the disorder of the free energy F,

$$E(F) = E(-kT \ln Z).\tag{4}$$

This calculation can rarely be done exactly. The determination of the probability distribution of F is an even more difficult task. One of the very few cases for which such a calculation can be done exactly is the Random Energy Model [13]. It is therefore highly desirable to investigate other explicitly solvable cases [7, 23], where the usual tools of disordered systems such as replica methods and variational techniques can be tested [7]. Other functionals of the Brownian motion which occur in the context of one-dimensional Anderson localization have also been studied in [22].

2. Physical Motivation: A Toy-model for Disordered Systems

Let us consider a particle confined on the interval $0 \le x \le L$ and submitted to a random force $F(x)$ distributed as a Gaussian white noise around some mean value $-f_0$. The corresponding random potential is then simply a Brownian motion with drift which can be written in terms of the Wiener process B_x as follows:

$$U(x) = -\int_0^x F(y)\, \mathrm{d}y \overset{\text{def}}{=} f_0 x + \sqrt{\sigma} B_x.\tag{5}$$

For a given sample, that is for a given realization of the potential $U(x)$, we define the partition function

$$Z_L = \int_0^L dx\, e^{-\beta U(x)}, \tag{6}$$

where as usual β is the inverse temperature of the system. It is convenient to introduce $\alpha = \beta^2 \sigma/2$ and the dimensionless parameter $\mu = 2f_0/\beta\sigma$ in order to rewrite Z_L as

$$Z_L^{(\mu)} = \int_0^L dx\, \exp[-(\alpha\mu x + \sqrt{2\alpha}\, B_x)]. \tag{7}$$

Therefore, for $\alpha = 2$, $Z_L^{(\mu)}$ and $A_L^{(\mu)}$ coincide. For $\alpha \neq 2$, using the scaling properties of the Brownian motion, one obtains

$$Z_L^{(\mu)} \overset{\text{(law)}}{=} \frac{2}{\alpha} \int_0^{\alpha L/2} dx\, \exp[-2(\mu x + B_x)]. \tag{8}$$

Hence

$$Z_L^{(\mu)} \overset{\text{(law)}}{=} \frac{2}{\alpha} A_{\alpha L/2}^{(\mu)}. \tag{9}$$

The thermal average of any function $g(x)$ of the position of the particle for a given sample, will be denoted by an upper-bar, i.e.

$$\overline{g(x)} \equiv \frac{\int_0^L dx\, g(x)e^{-\beta U(x)}}{\int_0^L dx\, e^{-\beta U(x)}}. \tag{10}$$

For instance, the thermal average and variance of the position are written as follows

$$\overline{x} = \frac{1}{Z_L^{(\mu)}} \left(-\frac{1}{\alpha} \frac{\partial}{\partial \mu} Z_L^{(\mu)} \right) = -\frac{1}{\alpha} \frac{\partial}{\partial \mu} \ln(Z_L^{(\mu)}), \tag{11}$$

$$\overline{x^2} - (\overline{x})^2 = \frac{1}{\alpha^2} \frac{\partial^2}{\partial \mu^2} \ln(Z_L^{(\mu)}). \tag{12}$$

More generally, the generating function of the thermal cumulants of the position reads

$$\ln(\overline{\exp[-px]}) = \ln Z_L^{(\mu+p/\alpha)} - \ln Z_L^{(\mu)}. \tag{13}$$

These relations show that the statistical properties of the position of the particle in the case $\mu = 0$ require, in fact, the knowledge of the partition function with an arbitrary drift μ.

The fundamental quantity for the statistical mechanics of this disordered system is the free energy $F_L^{(\mu)}$ related to the logarithm of the partition function,

$$F_L^{(\mu)} = -\frac{1}{\beta} \ln Z_L^{(\mu)}. \tag{14}$$

This paper deals essentially with the statistical properties of the free energy, and particularly with the mean free energy over the disorder, denoted by

$$E(F_L^{(\mu)}) = -\frac{1}{\beta} E(\ln Z_L^{(\mu)}). \tag{15}$$

This work also gives us the opportunity to bring together various results which were scattered throughout the physics and mathematics literature. A comparison between them often yields some interesting identities with a non-trivial probabilistic content.

The paper is organized as follows. In Section 3, we consider the case of random potential with zero drift $\mu = 0$. We first give the probability distribution of the free energy. We then establish various formulae for its mean value, using in particular the Bougerol identity [2, 6], and show that the replica method gives the correct result. We also discuss the asymptotic behavior of the mean free energy in the limit of a large sample $L \to \infty$. In Section 4, we discuss the properties of the free energy in the case of a random potential with a positive drift $\mu > 0$. We also establish some relations between $E(\ln Z_L^{(\mu)})$ and $E(1/Z_L^{(\nu)})$. In Section 5, we discuss the case where the length of the sample is an independent random variable which is exponentially distributed.

For the notation and properties of the special functions appearing in this paper, we refer the reader to Lebedev [20].

3. Case of Random Potential with Zero Drift $\mu = 0$

3.1. Distribution of the Partition Function $Z_L^{(0)}$ and Associated Free Energy

The expression of the generating function of $Z_L^{(0)}$ has already been derived in another context, through the resummation of the series of moments [24] and by a path-integral approach [21],

$$E(e^{-pZ_L^{(0)}}) = \frac{2}{\pi} \int_0^\infty dk \ \cosh \frac{\pi k}{2} K_{ik}\left(2\sqrt{\frac{p}{\alpha}}\right) \exp\left[-k^2 \frac{\alpha L}{4}\right]. \tag{16}$$

We also refer the reader to [1], where it is shown that this result may be derived from the Bougerol formula (see equation (31)). To invert the Laplace transform (16), it is convenient to start from the following integral representation,

$$K_{ik}\left(2\sqrt{\frac{p}{\alpha}}\right) = \int_0^\infty dt \ \cos kt \exp\left[-2\sqrt{\frac{p}{\alpha}} \cosh t\right]; \tag{17}$$

and to perform the integration over k in (16) to obtain the following:

$$E(e^{-pZ_L^{(0)}})$$

$$= \frac{2}{\sqrt{\pi\alpha L}} \exp\left[\frac{\pi^2}{4\alpha L}\right] \int_0^\infty dt \cos\left(\frac{\pi t}{\alpha L}\right) \exp\left[-\frac{t^2}{\alpha L}\right] \exp\left[-2\sqrt{\frac{p}{\alpha}}\cosh t\right].$$

(18)

We may then use the following identity,

$$\exp\left[-2\sqrt{\frac{p}{\alpha}}\cosh t\right] = \frac{\cosh t}{\sqrt{\pi\alpha}} \int_0^\infty \frac{dZ}{Z^{3/2}} e^{-pZ} \exp\left[-\frac{\cosh^2 t}{\alpha Z}\right],$$

(19)

to cast (18) into the form

$$E(e^{-pZ_L^{(0)}}) = \int_0^\infty dZ e^{-pZ} \psi_L^{(0)}(Z),$$

(20)

where

$$\psi_L^{(0)}(Z) = \frac{2\exp\left[\pi^2/4\alpha L\right]}{\pi\alpha\sqrt{L}} \frac{1}{Z^{3/2}} \int_0^\infty dt \cosh t \cos\left(\frac{\pi t}{\alpha L}\right) \ldots$$

$$\ldots \exp\left[-\frac{t^2}{\alpha L}\right] \exp\left[-\frac{\cosh^2 t}{\alpha Z}\right]$$

(21)

denotes the probability distribution of the partition function $Z_L^{(0)}$. A simple change of variables then gives the probability distribution $P_L^{(0)}(F)$ of the free energy $F_L^{(0)} = -(1/\beta)\ln Z_L^{(0)}$ as follows,

$$P_L^{(0)}(F) = \frac{2\beta\exp[\pi^2/4\alpha L]}{\pi\alpha\sqrt{L}} \exp\left[\frac{\beta}{2}F\right] \ldots$$

$$\ldots \int_0^\infty dt \cosh t \cos\left(\frac{\pi t}{\alpha L}\right) \exp\left[-\frac{t^2}{\alpha L}\right] \exp\left[-\left(\frac{\cosh^2 t}{\alpha}\right)e^{\beta F}\right].$$

(22)

It is interesting to compare (22) with a similar formula given in [7, Equation (8)]. Although the two expressions are very similar, they do not exactly coincide. The discrepancy comes from a different choice of the boundary conditions satisfied by the random potential $U(x)$ given in (5).

For large L, the probability density $X_L(\xi)$ of the reduced variable $\xi = (\beta F - \ln\alpha)/\sqrt{2\alpha L}$ tends to the Gaussian,

$$X_L(\xi) \xrightarrow[L\to\infty]{} \frac{1}{\sqrt{2\pi}} e^{-\xi^2/2}.$$

(23)

This asymptotic result may in fact be directly obtained since from the definition (7),

$$Z_L^{(0)} = \int_0^L dx\, e^{\sqrt{2\alpha}B_x} \overset{(\text{law})}{=} L\int_0^1 ds\, e^{\sqrt{2\alpha L}B_s},$$

(24)

hence

$$\frac{1}{\sqrt{L}} \ln Z_L^{(0)} = \frac{\ln L}{\sqrt{L}} + \ln \left(\int_0^1 ds \ e^{\sqrt{2\alpha L} \ B_s} \right)^{1/\sqrt{L}}. \tag{25}$$

Since

$$\ln \left(\int_0^1 ds \ e^{\sqrt{2\alpha L} \ B_s} \right)^{1/\sqrt{L}} \xrightarrow[L \to \infty]{} \ln \ \exp \left[\sqrt{2\alpha} \ \sup_{s \leq 1} B_s \right] \stackrel{\text{(law)}}{=} \sqrt{2\alpha} \ |N|, \tag{26}$$

where N is a normalized Gaussian variable, one has

$$\frac{1}{\sqrt{2\alpha L}} \ln Z_L^{(0)} \xrightarrow[L \to \infty]{\text{(law)}} |N|. \tag{27}$$

3.2. An Expression of $E\big(\ln Z_L^{(0)}\big)$ Derived from the Generating Function

The Frullani identity,

$$\ln Z_L^{(0)} = \int_0^\infty \frac{dp}{p} [e^{-p} - e^{-pZ_L^{(0)}}], \tag{28}$$

may be used to compute the mean of the logarithm of $Z_L^{(0)}$ from the generating function of equation (16). Using the intermediate regularization,

$$E(\ln Z_L^{(0)}) = \lim_{\epsilon \to 0+} \left[\Gamma(\epsilon) - \int_0^\infty dp \ p^{\epsilon - 1} \ E(e^{-pZ_L^{(0)}}) \right], \tag{29}$$

one obtains

$$E(\ln Z_L^{(0)}) = \frac{2}{\pi} \int_0^\infty \frac{dk}{k^2} [1 - e^{-\alpha L k^2} \pi k \coth(\pi k)] + C - \ln \alpha, \tag{30}$$

where $C = -\Gamma'(1)$ is Euler's constant.

3.3. Bougerol's Identity

Bougerol's identity [6] is an identity in law relating two independent Brownian motions $(B_s, s \geq 0)$ and $(\gamma_u, u \geq 0)$, and involving the exponential functional we are interested in. The statement is that for fixed t,

$$\sinh(B_t) \stackrel{\text{(law)}}{=} \gamma_{A_t} \quad \text{where} \quad A_t = \int_0^t ds \ e^{-2B_s} (= A_t^{(0)}.). \tag{31}$$

In the Appendix, we present a simple proof of this identity due to Alili, and Dufresne; we refer the reader to [1] for further details and possible generalizations for the case of a non-vanishing drift, $\mu \neq 0$.

Scaling properties of Brownian motion then give

$$A_t \overset{(\text{law})}{=} t \int_0^1 du \, e^{-2\sqrt{t}\, B_u}, \tag{32}$$

$$\sinh(B_t) \overset{(\text{law})}{=} \left(\int_0^1 du \, e^{-2\sqrt{t}\, B_u} \right)^{1/2} \gamma_t \tag{33}$$

$$\overset{(\text{law})}{=} \left(\frac{A_t}{t} \right)^{1/2} \gamma_t. \tag{34}$$

It follows that

$$E(\ln A_t) = E\left(\ln \left(\frac{\sinh(B_t)}{B_t} \right)^2 \right) + \ln t$$

$$= \int_{-\infty}^{+\infty} \frac{dx}{\sqrt{2\pi t}} \exp\left[-\frac{x^2}{2t} \right] \ln \left(\frac{\sinh(x)}{x} \right)^2 + \ln t. \tag{35}$$

Starting with the partition function

$$Z_L^{(0)} \overset{(\text{law})}{=} \frac{2}{\alpha} A_{\alpha L/2}, \tag{36}$$

we thus get a new expression for the mean value of the logarithm,

$$E(\ln Z_L^{(0)}) = \int_{-\infty}^{+\infty} \frac{dx}{\sqrt{\pi \alpha L}} \exp\left[-\frac{x^2}{\alpha L} \right] \ln \left(\frac{\sinh(x)}{x} \right)^2 + \ln(L). \tag{37}$$

Comparison with our previous result (30) leads to the identity,

$$2 \int_{-\infty}^{+\infty} \frac{dx}{\sqrt{\pi \alpha L}} \exp\left[-\frac{x^2}{\alpha L} \right] \ln \left(\frac{\sinh(x)}{x} \right)$$

$$= \frac{2}{\pi} \int_{-\infty}^{+\infty} \frac{dk}{k^2} \left[1 - e^{-\alpha L k^2} \pi k \coth(\pi k) \right] + C - \ln(\alpha L). \tag{38}$$

In order to understand the meaning of this identity, we differentiate both sides with respect to L and use the heat equation,

$$\frac{\partial}{\partial L} \left(\frac{1}{\sqrt{\pi \alpha L}} \exp\left[-\frac{x^2}{\alpha L} \right] \right) = \frac{\alpha}{4} \frac{\partial^2}{\partial^2 x} \left(\frac{1}{\sqrt{\pi \alpha L}} \exp\left[-\frac{x^2}{\alpha L} \right] \right). \tag{39}$$

After an easy integration by parts we obtain,

$$\int_{-\infty}^{+\infty} \frac{dx}{\sqrt{\pi \alpha L}} \exp\left[-\frac{x^2}{\alpha L} \right] \left[\frac{\partial^2}{\partial^2 x} \ln \left(\frac{\sinh(x)}{x} \right) \right]$$

$$= \alpha \int_{-\infty}^{+\infty} dk \, e^{-\alpha L k^2} \left[k \coth(\pi k) - |k| \right]. \tag{40}$$

This identity is a particular instance of Plancherel's formula, since the Fourier transform of the function

$$\frac{\partial^2}{\partial^2 x} \ln \left(\frac{\sinh(x)}{x} \right) = \frac{1}{x^2} - \frac{1}{\sinh^2(x)}, \tag{41}$$

can be easily obtained after an integration in the complex plane,

$$\int_{-\infty}^{+\infty} dx \, e^{ikx} \left[\frac{1}{x^2} - \frac{1}{\sinh^2(x)} \right] = \pi \left[k \coth \left(\frac{\pi}{2} k \right) - |k| \right]. \tag{42}$$

This formula is also encountered as follows in the study [3] of the Hilbert transform of Brownian motion

$$H_u = \lim_{\alpha \to 0^+} \int_0^u \frac{ds}{B_s} 1_{(|B_s| \geq \alpha)}. \tag{43}$$

If $\mathbf{n}(d\epsilon)$ denotes the characteristic measure of Brownian excursions, and ϵ the generic excursion with lifetime $V(\epsilon) = \inf\{t > 0 : \epsilon(t) = 0\}$, then the law of $H_V \equiv \int_0^V ds/\epsilon_s$ under $\mathbf{n}$ is dx/x^2, and (42) may be interpreted as follows,

$$\pi \left(\lambda \coth \left(\frac{\pi \lambda}{\theta} \right) - |\lambda| \right) = \int_{-\infty}^{+\infty} \frac{dx}{x^2} e^{i\lambda x} \, \mathbf{n} \left(1 - \exp \left[\frac{-\theta^2 V}{2} \right] \middle| H_V = x \right), \tag{44}$$

with

$$\mathbf{n} \left(1 - \exp \left[\frac{-\theta^2 V}{2} \right] \middle| H_V = x \right) = 1 - \left(\frac{\theta x/2}{\sinh(\theta x/2)} \right)^2, \tag{45}$$

in the particular case $\theta = 2$.

3.4. Mean Free Energy through the Replica Method

Let $X = Z_L^{(0)}$. The replica method is based on the identity,

$$E[\ln X] = \lim_{n \to 0} \frac{E[X^n] - 1}{n}. \tag{46}$$

In many applications, one proceeds by looking for an analytic continuation $n \to 0$ of the expressions of integer moments of X. This procedure is in general mathematically ill-founded, but in the present case non-integer moments can be computed.

The integer moments of the following partition function are well known [12, 21, 24, 31],

$$E[(Z_L^{(0)})^n] = \frac{1}{\alpha^n} \sum_{k=1}^{n} e^{\alpha L k^2} (-1)^{n-k} \frac{\Gamma(n)}{\Gamma(2n)} C_n^k + \frac{(-1)^n}{n!}. \tag{47}$$

Since this expression contains a sum of n terms, it has no *a priori* meaning for non-integer n. However, the following integral representation for the moment of order n [24],

$$E[(Z_L^{(0)})^n] = 2\frac{\Gamma(\frac{1}{2})}{\Gamma(n+\frac{1}{2})}\int_0^\infty \frac{dx}{\sqrt{\pi\alpha L}}\exp\left[-\frac{x^2}{\alpha L}\right]\left(\frac{\sinh^2 x}{\alpha}\right)^n, \qquad (48)$$

is valid for any positive real $n \geq 0$, as can be shown using the consequence (34) of Bougerol's identity (31).

The following expansions in n, as $n \to 0$,

$$\left(\frac{\sinh^2 x}{\alpha}\right)^n = 1 + n\ln\left(\frac{\sinh^2 x}{\alpha}\right) + o(n), \qquad (49)$$

$$\frac{\Gamma(\frac{1}{2})}{\Gamma(n+\frac{1}{2})} = 1 - n\frac{\Gamma'(\frac{1}{2})}{\Gamma(\frac{1}{2})} + o(n) = 1 + n(C + 2\ln 2) + o(n), \qquad (50)$$

where $C = -\Gamma'(1)$ denotes the Euler constant, lead to the expression

$$E[\ln(Z_L^{(0)})] = \lim_{n\to 0}\frac{E((Z_L^{(0)})^n) - 1}{n} \qquad (51)$$

$$= C + 2\ln 2 - \ln\alpha + \int_{-\infty}^{+\infty}\frac{dx}{\sqrt{\pi\alpha L}}\exp\left[-\frac{x^2}{\alpha L}\right]\ln(\sinh^2 x) \quad (52)$$

This formula is of course directly related to the previous expression (37) obtained using Bougerol's identity, since

$$-\int_{-\infty}^{+\infty}\frac{dx}{\sqrt{\pi\alpha L}}\exp\left[-\frac{x^2}{\alpha L}\right]\ln(x^2) + \ln(L) = C + 2\ln 2 - \ln\alpha. \qquad (53)$$

3.5.   Asymptotic Expression of the Mean Free Energy for Large L

The various expressions obtained previously (30), (37), (52) yield the following asymptotic behavior:

$$E[\ln(Z_L^{(0)})] = 2\sqrt{\frac{\alpha L}{\pi}} + C - \ln\alpha - \frac{\pi^{3/2}}{3\sqrt{\pi\alpha L}} + O\left(\frac{1}{(\alpha L)^{3/2}}\right) \quad \text{as } L \to \infty \qquad (54)$$

which agrees with [7].

We now show how the first two terms may in fact be recovered through a direct asymptotic analysis of

$$Z_L^{(0)} = \int_0^L dx\, e^{\sqrt{2\alpha}B_x}. \qquad (55)$$

Using the scaling properties of Brownian motion, one has

$$Z_L^{(0)} \stackrel{\text{(law)}}{=} \frac{1}{2\alpha} \int_0^{2\alpha L} \mathrm{d}x \, e^{B_x}. \tag{56}$$

It is convenient to set $\Lambda = 2\alpha L$ and to define

$$I_\Lambda = E\left[\ln \int_0^\Lambda \mathrm{d}x \, e^{B_x}\right]. \tag{57}$$

Introducing the one-sided supremum $S_\Lambda = \sup_{x \leq \Lambda} B_x$, one has

$$I_\Lambda = E(S_\Lambda) + E\left[\ln \int_0^\Lambda \mathrm{d}x \exp[-(S_\Lambda - B_x)]\right]. \tag{58}$$

It follows from the scaling properties of Brownian motion that

$$I_\Lambda = \sqrt{\Lambda}E(S_1) + E\left[\ln\left(\Lambda \int_0^1 \mathrm{d}x \exp[-\sqrt{\Lambda}(S_1 - B_x)]\right)\right]. \tag{59}$$

Recall that, from the reflection principle, $S_1 \stackrel{\text{(law)}}{=} |B_1|$. This implies the following:

$$E(S_1) = E(|B_1|) = \sqrt{\frac{2}{\pi}}. \tag{60}$$

For the remaining part of I_Λ, [26, 27, Theorem 1] gives the convergence in law, i.e.

$$\Lambda \int_0^1 \mathrm{d}x \exp[-\sqrt{\Lambda}(S_1 - B_x)] \xrightarrow[\Lambda \to \infty]{\text{(law)}} 4(T_1 + \hat{T}_1), \tag{61}$$

where T_1 and $\hat{T}_1$ are two independent copies of the first hitting time of 1 by a two-dimensional Bessel process starting from 0. Hence

$$E\left[\ln\left(\Lambda \int_0^1 \mathrm{d}x \exp[-\sqrt{\Lambda}(S_1 - B_x)]\right)\right] \xrightarrow[\Lambda \to \infty]{} E[\ln 4(T_1 + \hat{T}_1)]. \tag{62}$$

The right-hand side of the last equation can be evaluated either by direct calculation or by appealing to recent results related to the 'agreement formula' for Bessel processes [28]. The direct calculation proceeds as follows: one starts with the identity

$$E[\ln(T_1 + \hat{T}_1)] = \int_0^\infty \frac{\mathrm{d}u}{u}[e^{-u} - E(e^{-u(T_1 + \hat{T}_1)})], \tag{63}$$

then uses the expression of the Laplace transform [18],

$$E(e^{-uT_1}) = \frac{1}{I_0(\sqrt{2u})}.$$ (64)

After an intermediate regularization, one obtains

$$E[\ln(T_1 + \hat{T}_1)] = \lim_{\epsilon \to 0^+} \int_\epsilon^\infty \frac{du}{u} \left[e^{-u} - \frac{1}{I_0^2(\sqrt{2u})} \right]$$ (65)

$$= \lim_{\epsilon \to 0^+} \left[-C - \ln \frac{\epsilon^2}{2} - 2\frac{K_0(\epsilon)}{I_0(\epsilon)} \right] = C - \ln 2.$$ (66)

The other method relies on the following consequence of the 'agreement formula' for Bessel processes [27, equation (37)],

$$E_\delta[(T_1 + \hat{T}_1)^{(\delta/2)-1}] = \frac{1}{2^\mu \Gamma(1 + \mu)}.$$ (67)

Here $\delta = 2(1 + \mu)$ is the dimension of the Bessel process, and E_δ denotes the expectation with respect to the law of this process, starting at 0. (For $\delta = 2$, or equivalently $\mu = 0$, we simply write E instead of E_2.) By differentiating both sides of (67) with respect to μ at $\mu = 0$, one recovers (66):

$$E[\ln(T_1 + \hat{T}_1)] = C - \ln 2.$$ (68)

Finally, combining (62) and (68), we obtain

$$I_\Lambda - \sqrt{\frac{2\Lambda}{\pi}} \xrightarrow[\Lambda \to \infty]{} C + \ln 2.$$ (69)

We therefore recover the first two terms of (54)

$$E[\ln(Z_L^{(0)})] - \left(2\sqrt{\frac{\alpha L}{\pi}} + C - \ln \alpha \right) \xrightarrow[L \to \infty]{} 0.$$ (70)

4.   Case of Random Potential with Drift $\mu > 0$

4.1.   Distribution of the Partition Function $Z_L^{(\mu)}$ and Associated Free Energy

The probability distribution $\psi_L^{(\mu)}(Z)$ of the partition function $Z_L^{(\mu)}$ has already been obtained in another context, as the solution of the Fokker–Planck equation [11, 21]:

$$\frac{\partial \psi_L^{(\mu)}(Z)}{\partial L} = \frac{\partial}{\partial Z} \left[\alpha Z^2 \frac{\partial \psi_L^{(\mu)}(Z)}{\partial Z} + ((\mu + 1)\alpha Z - 1)\psi_L^{(\mu)}(Z) \right],$$ (71)

satisfying the initial condition $\psi_{L=0}^{(\mu)}(Z) = \delta(Z)$. The eigenfunction expansion of the solution exhibits the following relaxation spectrum with the length L,

$$\psi_L^{(\mu)}(Z) = \alpha \sum_{0 \le n < \mu/2} \exp[-\alpha L n (\mu - n)] \frac{(-1)^n (\mu - 2n)}{\Gamma(1 + \mu - n)} \left(\frac{1}{\alpha Z}\right)^{1+\mu-n} \cdots$$

$$\cdots L_n^{\mu-2n}\left(\frac{1}{\alpha Z}\right) \exp\left[-\frac{1}{\alpha Z}\right] \tag{72}$$

$$+ \frac{\alpha}{4\pi^2} \int_0^\infty ds \exp\left[-\frac{\alpha L}{4}(\mu^2 + s^2)\right] s \sinh \pi s \left|\Gamma\left(-\frac{\mu}{2} + i\frac{s}{2}\right)\right|^2 \left(\frac{1}{\alpha Z}\right)^{(1+\mu)/2} \cdots$$

$$\cdots W_{(1+\mu)/2,i(s/2)}\left(\frac{1}{\alpha Z}\right) \exp\left[-\frac{1}{2\alpha Z}\right], \tag{73}$$

where L_n^α are Laguerre's polynomials and $W_{p,\nu}$ are Whittaker's functions. Contrary to the case $\mu = 0$, there exists, for $\mu > 0$, a limit distribution as $L \to \infty$, i.e.

$$\psi_\infty^{(\mu)}(Z) = \frac{\alpha}{\Gamma(\mu)} \left(\frac{1}{\alpha Z}\right)^{1+\mu} \exp\left[-\frac{1}{\alpha Z}\right]. \tag{74}$$

In the physics literature, this limit distribution was first obtained in the context of one-dimensional Brownian diffusion in a Brownian drifted potential, where $Z_\infty^{(\mu)}$ plays the role of an effective trapping time [4], and was recently rediscovered in another context [23]. In the mathematics literature, the following limit distribution,

$$A_\infty^{(\mu)} \stackrel{(\text{law})}{=} \frac{1}{2Y_\mu}, \tag{75}$$

where Y_μ is a Gamma variable with parameter μ, and

$$P(Y_\mu \in (y, \ y + dy)) = \frac{dy}{\Gamma(\mu)} y^{\mu-1} e^{-y} \tag{76}$$

was first obtained by Dufresne [12], then shown in [34] to be another expression of the law of last passage times for transient Bessel processes. It is interesting to point out that (71) first appeared in [30] (see also [16]). However, the precise connection with our work is beyond the scope of this paper.

4.2. Mean Free Energy $E(\ln Z_L^{(\mu)})$

The Laplace transform of $Z_L^{(\mu)}$ corresponding to the probability distribution (72), (73) reads as follows [21],

$$E(e^{-pZ_L^{(\mu)}}) = \sum_{0 \leq n < \frac{\mu}{2}} \exp[-\alpha \, Ln(\mu-n)] \frac{2(\mu-2n)}{n!\Gamma(1+\mu-n)} \left(\frac{p}{\alpha}\right)^{\mu/2} K_{\mu-2n}\left(2\sqrt{\frac{p}{\alpha}}\right)$$

$$\tag{77}$$

$$+ \frac{1}{2\pi^2} \int_0^\infty ds \, \exp\left[-\frac{\alpha L}{4}(\mu^2 + s^2)\right] s \sinh \pi s \left|\Gamma\left(-\frac{\mu}{2} + i\frac{s}{2}\right)\right|^2$$

$$\times \left(\frac{p}{\alpha}\right)^{\mu/2} K_{is}\left(2\sqrt{\frac{p}{\alpha}}\right).$$

$$\tag{78}$$

We again use the Frullani identity

$$E(\ln Z_L^{(\mu)}) = \int_0^\infty \frac{dp}{p} \, [e^{-p} - E(e^{-pZ_L^{(\mu)}})],$$

$$\tag{79}$$

to obtain, after an intermediate regularisation (as in (29)),

$$E(\ln Z_L^{(\mu)}) = -\ln \alpha - \frac{\Gamma'(\mu)}{\Gamma(\mu)} - \sum_{1 \leq n < \mu/2} \frac{\mu - 2n}{n(\mu - n)} \exp[-\alpha Ln(\mu - n)]$$

$$\tag{80}$$

$$- 2 \int_0^\infty ds \frac{s}{\mu^2 + s^2} \frac{\sinh \pi s}{\cosh \pi s - \cos \pi \mu} \exp\left[-\frac{\alpha L}{4}(\mu^2 + s^2)\right].$$

$$\tag{81}$$

Obviously, the two first terms correspond to the contribution of the limit distribution (74),

$$E(\ln Z_\infty^{(\mu)}) = -\ln \alpha - \frac{\Gamma'(\mu)}{\Gamma(\mu)}.$$

$$\tag{82}$$

Now, it is easily deduced from (80), (81) and (82) that,

$$E(\ln Z_\infty^{(\mu)}) - E(\ln Z_L^{(\mu)}) \underset{L \to \infty}{\simeq} \begin{cases} \left(\frac{\mu-2}{\mu-1}\right) e^{-\alpha(\mu-1)L}, & \text{if } \mu > 2, \\[2mm] \frac{1}{\sqrt{\pi\alpha L}} e^{-\alpha L}, & \text{if } \mu = 2, \\[2mm] \frac{1}{\mu^2(1-\cos \pi\mu)} \left(\frac{\pi}{\alpha L}\right)^{3/2} e^{-\alpha\mu^2 L/4}, & \text{if } \mu < 2. \end{cases}$$

$$\tag{83}$$

It is interesting to compare these results to those obtained in [17] for the limiting distribution of the maximum of a diffusion process in a Brownian drifted environment. The latter may be stated as

$$
E\left(\frac{A_\infty^{(\mu)}}{A_\infty^{(\mu)} + \tilde{A}_L^{(-\mu)}}\right) \underset{L\to\infty}{\simeq}
\begin{cases}
\left(\frac{\mu-2}{\mu-1}\right) e^{-2(\mu-1)L}, & \text{if } \mu > 2, \\[2ex]
\frac{1}{\sqrt{2\pi L}} e^{-2L}, & \text{if } \mu = 2, \\[2ex]
\frac{K}{(2L)^{3/2}} e^{-\mu^2 L/2}, & \text{if } \mu < 2;
\end{cases}
\tag{84}
$$

where $A_\infty^{(\mu)}$ and $\tilde{A}_L^{(-\mu)}$ are two independent functionals of type $A_t^{(\nu)}$ defined in (1). The constant K is given in [17] as the following 4-fold integral,

$$
K = \frac{2^{1-\mu}}{\sqrt{\pi}} \frac{1}{\Gamma(\mu)} \int_0^\infty dy \; y^\mu \int_0^\infty dz \exp\left[-\frac{z}{2}(1+y^2)\right]
$$

$$
\times \int_0^\infty da \; a^{\mu-1} \frac{z}{z+a} e^{-a/2} \int_0^\infty du \; u \sinh u \exp[-zy\cosh u].
\tag{85}
$$

In fact (86) corrects the formula for the constant K found in [17]. The need to divide by $\sqrt{\pi}$ is due to a misprint in [31], where on page 528, just after formula (6.e), one should read $((2\pi^3 t)^{1/2})^{-1}\cdots$ instead of $((2\pi^2 t)^{1/2})^{-1}\cdots$.[1] The constant may in fact be explicitly computed (see below) to give the simple result,

$$
K = \pi^{3/2} \frac{\Gamma(\mu/2)^2}{\Gamma(\mu)} \frac{1}{(1 - \cos\pi\mu)}.
\tag{86}
$$

For $\mu \geq 2$, the result (84) therefore coincides with the result (83), where for the particular value $\alpha = 2$, $Z_L^{(\mu)}$ reduces to $A_L^{(\mu)}$ (see (7) and (1)). To understand this coincidence, we write the following:

$$
A_\infty^{(\mu)} = A_L^{(\mu)} + \exp[-2(B_L + \mu L)]\tilde{A}_\infty^{(\mu)},
\tag{87}
$$

where $\tilde{A}_\infty^{(\mu)}$ is a variable distributed as $A_\infty^{(\mu)}$ and independent of $A_L^{(\mu)}$. Therefore

$$
E(\ln A_\infty^{(\mu)}) - E(\ln A_L^{(\mu)}) = E\left(\ln\left[1 + \exp[-2(B_L + \mu L)]\frac{\tilde{A}_\infty^{(\mu)}}{A_L^{(\mu)}}\right]\right)
$$

$$
= E\left(\ln\left[1 + \frac{\tilde{A}_\infty^{(\mu)}}{A_L^{(-\mu)}}\right]\right).
\tag{88}
$$

The comparison between (83) and (84) therefore leads to

$$
E\left(\ln\left[1 + \frac{\tilde{A}_\infty^{(\mu)}}{A_L^{(-\mu)}}\right]\right) \underset{L\to\infty}{\simeq} E\left(\frac{\tilde{A}_\infty^{(\mu)}}{\tilde{A}_\infty^{(\mu)} + A_L^{(-\mu)}}\right) \qquad \text{for } \mu \geq 2.
\tag{89}
$$

---

[1] *Note added in Proof*: This has been corrected in paper [2], formula (6.e), in this book.

This is likely to be understood as a consequence of the following plausible statement,

$$\text{if } X_n \xrightarrow{\text{(a.s.)}} 0, \quad \text{then} \quad E[\ln(1 + X_n)] \sim E\left[\frac{X_n}{1 + X_n}\right]; \qquad (90)$$

which presumably holds for a large class of random variables $\{X_n\}$, but the precise conditions of validity of (90) elude us. However, (89) does not hold for $\mu < 2$ since, in this case, the prefactors in (83) and (84) differ.

To go further into the comparison, we have computed exactly the quantity occurring in (84) for arbitrary L. We start from the following identity:

$$E\left(\frac{Z_\infty^{(\mu)}}{Z_\infty^{(\mu)} + \tilde{Z}_L^{(-\mu)}}\right) = \int_0^\infty dp\, E(Z_\infty^{(\mu)} e^{-pZ_\infty^{(\mu)}})\, E(e^{-p\tilde{Z}_L^{(-\mu)}}). \qquad (91)$$

Using the following consequences of (78),

$$E(Z_\infty^{(\mu)} e^{-pZ_\infty^{(\mu)}}) = \frac{2}{\alpha\Gamma(\mu)}\left(\frac{p}{\alpha}\right)^{(\mu-1)/2} K_{\mu-1}\left(2\sqrt{\frac{p}{\alpha}}\right), \qquad (92)$$

and

$$E(e^{-p\tilde{Z}_L^{(-\mu)}}) = \frac{1}{2\pi^2}\int_0^\infty ds\, \exp\left[-\frac{\alpha L}{4}(\mu^2 + s^2)\right] s \sinh \pi s \left|\Gamma\left(\frac{\mu}{2} + i\frac{s}{2}\right)\right|^2 \cdots$$

$$\qquad (93)$$

$$\cdots \left(\frac{p}{\alpha}\right)^{-\mu/2} K_{is}\left(2\sqrt{\frac{p}{\alpha}}\right),$$

we get the following expression,

$$E\left(\frac{Z_\infty^{(\mu)}}{Z_\infty^{(\mu)} + \tilde{Z}_L^{(-\mu)}}\right) = \frac{1}{\pi^2\Gamma(\mu)}\int_0^\infty dx\, K_{\mu-1}(x)\cdots$$

$$\qquad (94)$$

$$\cdots \int_0^\infty ds\, \exp\left[-\frac{\alpha L}{4}(\mu^2 + s^2)\right] s \sinh \pi s \left|\Gamma\left(\frac{\mu}{2} + i\frac{s}{2}\right)\right|^2 K_{is}(x).$$

For $\mu \le 2$, the order of integrations may be exchanged to give,

$$E\left(\frac{Z_\infty^{(\mu)}}{Z_\infty^{(\mu)} + \tilde{Z}_L^{(-\mu)}}\right) = \frac{1}{2\Gamma(\mu)}\int_0^\infty ds\, \exp\left[-\frac{\alpha L}{4}(\mu^2 + s^2)\right]\cdots$$

$$\qquad (95)$$

$$\cdots \frac{s \sinh \pi s}{\cosh \pi s - \cos \pi\mu}\left|\Gamma\left(\frac{\mu}{2} + i\frac{s}{2}\right)\right|^2,$$

which reduces for $\mu = 2$ to

$$E\left(\frac{Z_\infty^{(2)}}{Z_\infty^{(2)} + \tilde{Z}_L^{(-2)}}\right) = \frac{1}{2}\int_0^\infty ds\, \exp\left[-\frac{\alpha L}{4}(4 + s^2)\right]\frac{\pi s^2 \cosh(\pi s/2)}{\sinh^2(\pi s/2)}. \qquad (96)$$

For $\mu > 2$, one cannot exchange the order of integrations in (94). However, one may start from a series representation [21, equation (5.9)] of the generating function (93) to obtain, after some algebra involving deformation of a contour integral in the complex plane (see [21] for a similar approach), the following general result for arbitrary $\mu > 0$;

$$
E\left(\frac{Z_\infty^{(\mu)}}{Z_\infty^{(\mu)} + \tilde{Z}_L^{(-\mu)}}\right)
$$

$$
= \sum_{1 \le n < \mu/2} (\mu - 2n)\frac{\Gamma(n)\Gamma(\mu - n)}{\Gamma(\mu)} \exp[-\alpha Ln(\mu - n)] \tag{97}
$$

$$
+ \frac{1}{2\Gamma(\mu)} \int_0^\infty ds \exp\left[-\frac{\alpha L}{4}(\mu^2 + s^2)\right] \frac{s \sinh \pi s}{\cosh \pi s - \cos \pi \mu} \left|\Gamma\left(\frac{\mu}{2} + i\frac{s}{2}\right)\right|^2. \tag{98}
$$

From (95), (96), (98), one easily recovers the corresponding asymptotic results of (84), and in particular the value of K given in (86), which were obtained in [17] by a quite different method, relying on the computations made in [31]. The presence of discrete terms for $\mu > 2$ again explains the transition at $\mu = 2$ of the asymptotic behavior.

4.3. Some Relations between $E(\ln Z_L^{(\mu)})$ and the Mean Inverse $E(1/Z_L^{(\nu)})$

The first negative moment can be obtained from the generating function (78) [21],

$$
E\left(\frac{1}{Z_L^{(\mu)}}\right) = \int_0^\infty dp \, E(e^{-p Z_L^{(\mu)}}) \tag{99}
$$

$$
= \alpha \sum_{0 \le n < \mu/2} (\mu - 2n) \exp[-\alpha Ln(\mu - n)]
$$

$$
+ \frac{\alpha}{2} \int_0^\infty ds \frac{s \sinh \pi s}{\cosh \pi s - \cos \pi \mu} \exp\left[-\frac{\alpha L}{4}(\mu^2 + s^2)\right]. \tag{100}
$$

The previous explicit expressions therefore lead to the very simple identity for any $\mu \ge 0$,

$$
\frac{\partial}{\partial L} E(\ln Z_L^{(\mu)}) = E\left(\frac{1}{Z_L^{(\mu)}}\right) - \alpha\mu. \tag{101}
$$

Can this be derived directly, using only basic properties of Brownian motion? Trying to do so reveals two other identities of the same kind, but not (101).

The first one relates the exponential functionals for two opposite drifts $(+\mu)$ and $(-\mu)$,

$$\frac{\partial}{\partial L} E(\ln Z_L^{(\mu)}) = E\left(\frac{1}{Z_L^{(-\mu)}}\right). \tag{102}$$

This identity can be obtained from a simple reparameterisation $x' = L - x$ in the denominator of the left-hand side of expression,

$$\frac{\partial}{\partial L} E\left(\ln \int_0^L dx \ \exp[-(\alpha\mu x + \sqrt{2\alpha}\ B_x)]\right)$$
$$= E\left(\frac{\exp[-(\alpha\mu L + \sqrt{2\alpha}\ B_L)]}{\int_0^L dx \exp[-(\alpha\mu x + \sqrt{2\alpha}\ B_x)]}\right). \tag{103}$$

The second identity relates the exponential functionals for two dimensionless drifts differing by 2,

$$\frac{\partial}{\partial L} E(\ln Z_L^{(\mu)}) = \exp[-\alpha L(\mu - 1)] \ E\left(\frac{1}{Z_L^{(\mu-2)}}\right). \tag{104}$$

This identity follows from Cameron–Martin or Girsanov relations. In fact these relations lead to a more general relation between the two characteristic functions of the exponential functionals for (μ) and $(\mu - 2)$,

$$\frac{\partial}{\partial L} E(e^{-pZ_L^{(\mu)}}) = -p \exp[-\alpha L(\mu - 1)] \ E(e^{-pZ_L^{(\mu-2)}}). \tag{105}$$

5.   Expressions of $E(\ln(Z_{L_\lambda}^{(\mu)}))$ with an Independent Exponential Length L_λ

It is well known that the laws of additive functionals of Brownian motion, with drift μ, i.e.

$$A_t^f \stackrel{(\mathrm{def})}{=} \int_0^t ds \ f(B_s + \mu s), \tag{106}$$

may be easier to compute when the fixed time t is replaced by an independent exponential time T_λ of parameter λ

$$P(T_\lambda \in [t, t + dt]) = \lambda \ e^{-\lambda t} dt. \tag{107}$$

This is indeed the case for $A_t^f = A_t^{(\mu)}$, i.e. $f(x) = e^{-2x}$. It was shown in [32, 33] that

$$A_{T_\lambda}^{(\mu)} \stackrel{(\mathrm{law})}{=} \frac{X_{1,a}}{2Y_b}, \quad \text{where } a = \frac{\sqrt{2\lambda + \mu^2} - \mu}{2}, \ b = \frac{\sqrt{2\lambda + \mu^2} + \mu}{2}, \tag{108}$$

where $X_{1,a}$ and Y_b are independent, $X_{\alpha,\beta}$ denotes a Beta-variable with parameters (α, β),

$$P(X_{\alpha,\beta} \in [x, x + dx]) = \frac{x^{\alpha-1}(1-x)^{\beta-1}}{B(\alpha, \beta)} dx, \quad (0 < x < 1), \quad (109)$$

and Y_γ denotes a Gamma-variable of parameter (γ) as in (76). Consequently, one has

$$E((A_{T_\lambda}^{(\mu)})^n) = \frac{\Gamma(1+n)\Gamma(1+a)\Gamma(b-n)}{2^n\Gamma(1+a+n)\Gamma(b)} \quad (110)$$

and

$$-E(\ln A_{T_\lambda}^{(\mu)}) = C + \ln 2 + \psi(1+a) + \psi(b), \quad (111)$$

where $C = -\Gamma'(1)$ is Euler's constant, and $\psi(x) = \Gamma'(x)/\Gamma(x)$. Classical integral representation of the function ψ allows to invert the Laplace transform in λ implicit in (111), hence to recover $E(\ln A_t^{(\mu)})$. However the formulae we have obtained in this way are not simple.

To transpose the result (111) for the partition function $Z_{L_\lambda}^{(\mu)}$, describing the case where the length of the disordered sample is exponentially distributed, one only needs to use the identity derived from (9):

$$Z_{L_\lambda}^{(\mu)} \overset{\text{(law)}}{=} \frac{2}{\alpha} A_{T_{(2/\alpha)\lambda}}^{(\mu)}. \quad (112)$$

Appendix.
A Simple Proof Bougerol's Identity

As Bougerol's identity (31) may appear quite mysterious at first sight, we find useful to reproduce here a simple proof of this identity due to L. Alili and D. Dufresne. We refer the reader to [1] and [2] for further details and possible generalizations for the case of a non-vanishing drift $\mu \neq 0$.

Consider the Markov process

$$X_t = e^{B_t} \int_0^t e^{-B_s} \, d\gamma_s, \tag{113}$$

where B_t and γ_t are two independent Brownian motions. The Itô formula yields the stochastic differential equation,

$$dX_t = \tfrac{1}{2} X_t \, dt + (X_t \, dB_t + d\gamma_t). \tag{114}$$

We introduce a new Brownian motion β_t, by setting

$$X_t \, dB_t + d\gamma_t = \sqrt{X_t^2 + 1} \, d\beta_t, \tag{115}$$

from which it follows that

$$dX_t = \tfrac{1}{2} X_t \, dt + (X_t^2 + 1)^{1/2} \, d\beta_t. \tag{116}$$

A comparison with

$$d[\sinh(\beta_t)] = \tfrac{1}{2} [\sinh(\beta_t)] \, dt + (\sinh^2(\beta_t) + 1)^{1/2} \, d\beta_t, \tag{117}$$

shows the identity in law between the following processes

$$(\sinh(\beta_t), t \geq 0) \overset{(\text{law})}{=} \left(X_t = e^{B_t} \int_0^t e^{-B_s} \, d\gamma_s; t \geq 0 \right) \tag{118}$$

whereas the stability of the law of Brownian motion by time reversal at fixed time t, gives

$$\sinh(\beta_t) \overset{(\text{law})}{=} \hat{\gamma}_{A_t^{(0)}} \quad \text{with} \quad A_t^{(0)} = \int_0^t e^{2B_s} \, ds, \quad \text{for fixed } t > 0, \tag{119}$$

where $\hat{\gamma}$ denotes a Brownian motion, which is independent of B.

Acknowledgements

We thank J.P. Bouchaud for interesting discussions at the beginning of this work, and B. Duplantier for a careful reading of the manuscript.

References

1. Alili, L. (1995). Fonctionnelles exponentielles et valeurs principales du mouvement Brownien. Thèse de l'Université Paris 6
2. Alili, L., Dufresne, D. and Yor, M. (1997). Sur l'identité de Bougerol pour les fonctionnelles exponentielles du mouvement Brownien avec drift. In *Exponential Functionals and Principal Values related to Brownian Motion. A collection of research papers; Biblioteca de la Revista Matematica, Ibero–Americana*, ed., M. Yor. p. 3–14.
3. Biane, P. and Yor, M. (1987). Valeurs principales associées aux temps locaux Browniens, *Bull. Sci. Math.*, **111**, 23–101
4. Bouchaud, J.P., Comtet, A., Georges, A. and Le Doussal, P. (1990). Classical diffusion of a particle in a one-dimensional random force field. *Ann. Phys.*, **201**, 285–341
5. Bouchaud, J.P. and Georges, A. (1990). Anomalous diffusion in disordered media: statistical mechanisms, models and physical applications. *Phys. Rep.*, **195**, 127–293
6. Bougerol, P. (1983). Exemples de théorèmes locaux sur les groupes résolubles. *Ann. Inst. H. Poincaré*, **19**, 369–391
7. Broderix, K. and Kree, R. (1995). Thermal equilibrium with the Wiener potential: testing the Replica variational approximation. *Europhys. Lett.*, **32**, 343–348
8. Burlatsky, S.F., Oshanin, G.H., Mogutov, A.V. and Moreau, M. (1992). Non-Fickian steady flux in a one-dimensional Sinaï-type disordered system. *Phys. Rev. A*, **45**, 6955–6957
9. Carmona, Ph., Petit, F. and Yor, M. (1994). Sur les fonctionnelles exponentielles de certains processus de Lévy. *Stochast. Rep.*, **47**, p. 71. **Paper [8] in this book**
10. Carmona, Ph., Petit, F. and Yor, M. (1997). On the distribution and asymptotic results for exponential functionals of Lévy processes. In *Exponential Functionals and Principal Values related to Brownian Motion. A collection of research papers Biblioteca de la Revista Matematica, Ibero–Americana*, ed., M. Yor. p. 73–126.
11. Comtet, A. and Monthus, C. (1996). Diffusion in one-dimensional random medium and hyperbolic Brownian motion. *J. Phys. A: Math. Gen.*, **29**, 1331–1345
12. Dufresne, D. (1990). The distribution of a perpetuity, with applications to risk theory and pension funding. *Scand. Act. J.*, 39–79
13. Gardner, E. and Derrida, B. (1989). The probability distribution of the partition function of the Random Energy Model. *J. Phys. A: Math. Gen.*, **22**, 1975–1981
14. Geman, H. and Yor, M. (1993). Bessel processes, Asian options and perpetuities. *Math. Fin.*, **3**, 349–375. **Paper [5] in this book**
15. Georges, A. (1988). Diffusion anormale dans les milieux désordonnés: Mécanismes statistiques, modèles théoriques et applications. Thèse d'état de l'Université Paris 11
16. Hongler, M.O. and Desai, R.C. (1986). Decay of unstable states in the presence of fluctuations, *Helv. Phys. Acta.*, **59**, 367–389
17. Kawazu, K. and Tanaka, H. (1993). On the maximum of a diffusion process in a drifted Brownian environment. *Sem. Prob.*, **XXVII**, Lect. Notes in Math. 1557, Springer, p. 78–85
18. Kent, J. (1978). Some probabilistic properties of Bessel functions. *Ann. Prob.*, **6**, 760–768

19. Kesten, H., Kozlov, M. and Spitzer, F. (1975). A limit law for random walk in a random environment. *Compositio Math.,* **30**, p. 145–168

20. Lebedev, N. (1972). *Special Functions and their Applications.* Dover

21. Monthus, C. and Comtet, A. (1994). On the flux in a one-dimensional disordered system. *J. Phys. I (France),* **4**, 635–653

22. Monthus, C., Oshanin, G., Comtet, A. and Burlatsky, S.F. (1996). Sample-size dependence of the ground-state energy in a one-dimensional localization problem. *Phys. Rev. E,* **54**, 231–242

23. Opper, M. (1993). Exact solution to a toy random field model. *J. Phys. A: Math. Gen.,* **26**, L719–L722

24. Oshanin, G., Mogutov, A. and Moreau, M. (1993). Steady flux in a continuous Sinaï chain. *J. Stat. Phys.,* **73**, 379–388

25. Oshanin, G., Burlatsky, S.F., Moreau, M. and Gaveau, B. (1993). Behavior of transport characteristics in several one-dimensional disordered systems. *Chem. Phys.,* **177**, 803–819

26. Pitman, J. and Yor, M. (1993a). A limit theorem for one-dimensional Brownian motion near its maximum, and its relation to a representation of the two-dimensional Bessel bridges. Preprint.

27. Pitman, J. and Yor, M. (1993b). Dilatations d'espace-temps, réarrangement des trajectoires Browniennes, et quelques extensions d'une identité de Knight, *C. R. Acad. Sci. Paris,* **316**, I 723–726

28. Pitman, J. and Yor, M. (1996). Decomposition at the maximum for excursions and bridges of one-dimensional diffusions. In *Itô's Stochastic Calculus and Probability Theory,* eds., N. Ikeda, S. Watanabe, M. Fukushima and H. Kunita. Springer, p. 293–310

29. De Schepper, A., Goovaerts, M. and Delbaen, F. (1992). The Laplace transform of annuities certain with exponential time distribution. *Ins. Math. Econ.,* **11**, p. 291–304

30. Wong, E. (1964). The construction of a class of stationary Markov processes. In *Am. Math. Soc. Proc. of the 16th Symposium of Appl. Math.* p. 264–285

31. Yor, M. (1992). On some exponential functionals of Brownian motion. *Adv. Appl. Prob.,* **24**, 509–531. **Paper [2] in this book**

32. Yor, M. (1992). Sur les lois des fonctionnelles exponentielles du mouvement Brownien, considérées en certains instants aléatoires. *C. R. Acad. Sci. Paris,* **314**, 951–956. **Paper [4] in this book**

33. Yor, M. (1992). Some aspects of Brownian motion. Part I: Some special functionals. *Lectures in Mathematics.* ETH Zürich, Birkhäuser

34. Yor, M. (1993). Sur certaines fonctionnelles exponentielles du mouvement Brownien réel. *J. Appl. Proba.,* **29**, 202–208. **Paper [1] in this book**

35. Yor, M. (1993). From planar Brownian windings to Asian options. *Ins. Math. Econ.,* **13**, 23–34. **Paper [7] in this book**

Postscript #10

a) This paper presents some motivations for the study of Brownian exponential functionals from questions arising in Physics, together with the relevant literature.

b) A number of formulae found in this paper are discussed and developed in *D. Dufresne* (2001): The integral of geometric Brownian motion. *Adv. App. Prob.*, **33**, 223–241 which contains a wealth of information about the law of $A_t^{(\mu)}$.

Index

American option, 1, 5, 81
Asian option, vii, 5, 9, 11, 23, 49, 106, 172

barrier options, 9
Bessel process, bridge, functions, vii, viii, 15, 20, 25, 65, 80, 86, 123, 125, 133
Bessel, viii
beta variable, 55
Black and Scholes formula, 76, 81
Black–Derman–Toy model, 106
Black-Scholes-Merton setting, model, formula, 3, 7, 79
Bougerol's identity, formula, 24, 25, 29, 34, 75, 103, 185, 187, 200
Brownian motion, vii, 23, 55, 96, 183

C.I.R, 82
Carleman's Criterion, 25, 74, 176
compound Poisson process, 140
confluent hypergeometric function, 52, 61, 78
Conformal Invariance, 112
Cox–Ingersoll–Ross model, 82, 83, 85, 87

Disordered Systems, 182
double-barrier options, 9, 11

Fellerian, 153
formula, 76
Excursion Theory, 127
exponential functionals, viii, 171

gamma variable, 15, 55, 82, 193
geometric Brownian motion, 1, 6, 93
Girsanov transformation, 141
Girsanov's relationship, 84
Girsanov's theorem, 3, 24, 39, 179

Hartman-Watson probability measure, distribution, 28, 42
hyperbolic Brownian motion, 24, 45
hyperbolic, 183

infinitesimal generator, 157
intertwined, 53
intertwining relation, 140
intertwinings, 141, 147

Lamperti relation, viii, 139, 182
Lamperti's representation, 171
last passage times, 20, 95, 193
Lévy processes, 139, 140, 141, 159
lognormal distribution, 63, 68, 74

Maximal equality, 17
model, 82
modified Bessel functions, 61, 93
moments, 27, 31, 74, 145
Monte Carlo simulations, 2, 8, 26, 68

Ornstein–Uhlenbeck processes, 63, 83, 172

Perpetuities, 63, 81, 172, 178
planar Brownian motion, 123
processes, 153

replica method, 185, 189
Riemann zeta function, 133

semigroup, 65, 125, 158
semigroups of Bessel processes, 51
semi-stable Markov processes, viii, 139, 147, 155, 157, 161
Skew-product Representation, 96, 112

winding number, 123

Printing (Computer to Film): Saladruck Berlin
Binding: Stürtz AG, Würzburg